BK
C4

Science is God

Science is God

David F Horrobin

MA, DPhil, BM, BCh,
Professor of Medical Physiology,
University College, Nairobi
Late Scholar of Balliol College and
Fellow of Magdalen College, Oxford

MTP

Medical and Technical Publishing Co Ltd
Chiltern House, Aylesbury

Published in Great Britain in 1969 by
MTP Chiltern House Aylesbury Bucks

SBN 85200 000 6

Printed in Great Britain by
Billing & Sons Limited
Guildford and London

Preface

I am becoming increasingly disturbed by the lack of understanding of science revealed by politicians, industrialists and the
general public. I am also concerned about the widespread misuse of the word "scientific" which is more and more being used
in situations where it is quite inappropriate. As a result, in some
circumstances gross overestimates are made as to what science
can do. In other circumstances the real power of science is
foolishly underestimated and the contributions which it can
make are squandered. *Science is God* is an attempt to explain
just what is meant by the scientific approach and to define more
closely what the word "scientific" indicates. It is deliberately
brief and controversial because I want it to be read. In fact, the
material dealt with in each single chapter really deserves a whole
book to itself. In the future I hope that I may be able to give to
each subject such full treatment. Meanwhile I hope that this
book will stimulate discussion about science and will increase
understanding of it.

DAVID F. HORROBIN
Nairobi, Kenya

For Nafisa and Cathra

Contents

Chapter 1

Introduction

Science is the most influential and the least understood of the forces shaping our world. In one form or another the cliché is piously repeated a million times a day by politicians, theologians and amateur philosophers. Yet little effective is done about it. This book is an attempt to explain why.

Scientists are neither saintly philosophers nor black-hearted devils. For the most part they are ordinary, simple, godless people whose only memorial will be a thousand lost golf balls. Like farmers and factory workers, bookmakers and businessmen, their main aim in life is to acquire enough money to provide security and pleasure for themselves, their wives and their children. And like most other people, scientists think of security and pleasure in the crude terms of income and expenditure over the coming year or two. No scientist I know loses much sleep over the effect his work is going to have on society: many spend lonely nights worrying about the effect their work is going to have on their own economic future. Armchair pundits may regard this as something reprehensible. Perhaps it is if to be human is reprehensible. Professional politicians, theologians, sociologists and philosophers may spend their lives worrying about society. However, their worrying functions are performed largely as a means to the end of personal ambition. When they lie restlessly awake, it is their professional future rather than the fate of society which concerns them. The unconcern of most scientists about so-called great issues indicates only that they are human beings and not that they are members of the criminal classes. In this respect their behaviour is no worse and no better than that of any other profession. However, because of the significance of science, this lemming unconcern may have dangerous consequences. It ensures that mankind remains ignorant of the forces shaping the world. This is hardly a new situation. Mankind always has been ignorant and probably always will be. We are merely unfortunate in that we live

9

in an age when these forces are driving us harder and faster than ever before. Nevertheless, like all humanity before us we shall probably continue to ignore them until they impinge upon the comfort of ourselves and our children.

If the Blake-like picture of scientists as angels or devils is misleading, so is the one of the scientist who understands all science. I am a physiologist, a specialised form of medical scientist who studies the normal functioning of the body. I understand only in the haziest way most other forms of medical science. I take all the influential general scientific journals and I pride myself on being an enlightened professional who tries to keep up with advances in as many fields as possible. Yet even in biology there are few topics whose full significance I appreciate. My knowledge of other sciences, of chemistry or physics, of astronomy or geology, is rudimentary. Each scientist understands fully only the tiniest fraction of the total range of scientific endeavour, and the better the scientist the greater is his awareness of his ignorance. The physiologist finds it as difficult to speak to the physicist in terms which both understand as either one of them does to the specialist in English literature. There are a thousand scientific cultures with a myriad of subcultures. There is no such thing as the scientific view of a problem. There is not even a chemist's view, a biologist's view or a physicist's view. Dr. Joe Bloggs of Utopia University presumes too much when he puts forward the views of medical scientists: what he means is the view of Dr. Joe Bloggs.

The failure of non-scientists to understand any science is a real problem. In answer to it some have advocated the mass education of general scientists to act as envoys between the two worlds. These men with a grasshopper knowledge of all scientific fields are supposed to go into industry and government to act as scientific advisers. All that this scheme will do is produce an array of fast talkers with misleading qualifications who, far from understanding all science, understand none properly. A scientist is a scientist, and a scientist is one who knows what it is like to carry out scientific research. Only a man who has grappled with them can understand the real problems in any scientific field. A man who has gone through the charade of attending a series of lectures and seminars in general science is not a scientist. He is no more likely to be able to understand scientific problems than one who has a degree in history.

Even more fatuous is the idea of providing crash courses for civil servants, businessmen and politicians. These concentrate on the "facts of science" and are therefore valueless. Since a highly trained scientist cannot fully understand the recent advances made even in a field quite closely related to his own, it is useless to expect those without any scientific training to do so. People are certain to come away from such a course with a multitude of misconceptions which may well be more dangerous than total ignorance. If these emergency educational schemes are to be of any value at all they should not concentrate on the facts of science because these are changing every day. Instead they should discuss the fundamental assumptions on which science is based and should try to explain how the scientific approach differs from all other approaches to a problem. They should attempt to give some idea of how in practice a scientist goes about his work and should make the businessman, the politician and the civil servant understand just what conditions the scientist needs.

Many authors have written about the facts of science in terms which the layman can understand. But masterly exposition of the facts of science does not explain the principal features of the scientific approach nor does it give any understanding of the way in which a scientist goes about scientific research. A much smaller group of writers, of whom Koestler[1] is the outstanding example, have attempted this much more difficult task of explaining to laymen just what is involved in the process of scientific discovery. Unfortunately, few of this second group are practising research scientists and few seem to appreciate what it is like to be involved in tackling an actual scientific problem. They tend to get themselves tied up in high-flown theories and, as Medawar has pointed out,[2] they sometimes seem to base their writing on the syllogism, "Profound thought is difficult to understand. My thought is difficult to understand. Therefore my thought is profound." Such obscure profundity is fashionable and most readers are taken in by it. The books are praised even

11

though they do not bring understanding, perhaps even *because* they do not bring understanding. Readers and reviewers have not the courage to admit that to them much of the work is incomprehensible nonsense. They fear that if they did admit this they might expose themselves to the charge of weakness of intellect. And so the emperors continue on their way, confident of the protection offered by their non-existent intellectual clothing. It is hardly surprising that they have made no visible impact on the problems of science and society.

Yet if laymen do not understand the scientific approach, scientists themselves can hardly blame either the laymen or those heroic souls I have just criticised who do at least attempt to make that approach intelligible to the scientifically un-sophisticated. It does not help to be rude if one is not prepared to tackle the problem oneself. Successful businessmen waste little time in pondering upon the effect of commerce on the future of the world or in studying the principles of economics. Similarly, successful scientists pay scant attention to the study of the scientific method or to the impact of science on society. Businessmen leave economic principles to academics whose relative poverty indicates their lack of understanding of the business world. Successful scientists leave the study of the scientific approach to philosophers who have nothing better to do, to once great scientists decaying into senility and to failed scientists turning to it as a last refuge. Since I am neither a philosopher nor a once great scientist, it will inevitably be assumed that I come into the last of these classes. I hope that this is not true, but it may be, for I have been so foolish as to leave the security of the Anglo-American academic scene for the uncertainties of Africa. But I hope, too, that this book, being more blunt and less obscure than others which have dealt with similar topics, may stimulate scientists to think a little more of things they have always taken for granted and may give laymen some faint understanding of what science is all about.

Chapter 2

The Assumptions of Science

Every scientist must make two assumptions which are quite unproveable, even in theory. The first is that the universe is orderly and the second is that man's brain is capable of unravelling the mysteries of that order. No scientist I know ever thinks about these assumptions, still less worries about them. They in no significant sense influence practical scientific activity. Nevertheless, they are made and they are worth examining briefly because the conclusions of science cannot be more reliable than the fundamental assumptions on which science is based.

The assumption that the universe is orderly is the central tenet of the scientific faith upon which all else depends. It is accepted without reserve by all scientific research workers. As a scientist I believe that the behaviour of the constituents of the universe can be described by certain rules known as scientific laws which vary neither in space nor in time. Even if the laws should turn out to vary in space or time, I believe that the variation must be orderly and must be capable of being described by another law. If I mix two chemicals together in England, I believe that the results of the reaction will be identical to those which would occur if I mixed the two together under similar conditions in Australia or in the United States. I even believe that the same results would be obtained in Russia and in Red China. These beliefs, of course, are not unreasonable and can be easily checked. I also believe that if I mix the two chemicals together this year, the results will be identical when I carry out the same procedure in a year's time. This, too, is a reasonable and verifiable belief. Having tested my faith in orderliness at a number of different times and in a number of geographical locations, I feel justified in summarising the results in terms of a law such as: "The reaction of sodium bicarbonate with hydrochloric acid yields sodium chloride and carbon dioxide." This law, I believe, will hold good for every second in time and at every point on the earth's surface. Naturally, I could never test the law at every second in time and at every point on the earth's surface. Strictly and pedantically speaking I could never prove its general application. However, if I test the law at several different times and places there can be no serious doubts as to its validity. I do not believe it very likely that there is some point on the earth where the law is untrue.

I carry my faith in the orderliness of the universe far beyond this. I believe that if I took the chemicals to Venus or even to some distant star and then mixed them under similar conditions, the results would be again identical to those I obtained on earth. This at the moment is unverifiable, yet I believe it without reserve. Every other scientist believes in the same way. Similarly, I believe that the results I obtain today are identical to the results I would have obtained had I carried out the experiment a million years ago. Again the belief is absolutely unverifiable and again I hold it without reserve.

I hold my unverifiable beliefs so tenaciously because I have found to my own satisfaction that they work within my own limited range of experience. They also work within the much wider range of experience of all other scientists. Furthermore, although my belief in the orderliness of the universe is, strictly speaking, unprovable, I can conceive of no circumstance which would lead me to reject it. If I found that my two chemicals did behave differently in Sydney, Australia, and in London, England, my faith would be in no way affected. I should not say that God had given a sign nor that a miracle had happened. I should merely assume that the unexpected behaviour could be accounted for by some law of which I was as yet unaware. If I were interested enough, and if I could raise the necessary financial support, I might then set about trying to discover the nature of this law. No doubt I would eventually arrive at some conclusion, satisfactory to myself if not to others. My belief in order would be vindicated.

In my more jaundiced moments I compare this behaviour to that of a priest of some ancient religion. This priest believed that all natural events could be accounted for by the activities of a group of Olympian gods. By careful study he came to understand in these terms the behaviour of sea and sky and earth and man. On being confronted by some strange event which he could not immediately explain, the priest did not desert his faith. After much thought and fasting the answer would come to him either in terms of the discovery of a new god or of an unusual combination of the old gods' activities. No doubt his conclusions were satisfactory both to himself and to his devotees.

I do not follow science rather than the mysteries of some ancient religion because the initial intellectual processes in the two cases are very different. I follow science merely because it has proved more powerful as a tool for controlling the human environment. The priest believed that the world was governed by law, but by law which was unpredictable in effect because it depended on the capricious activities of the gods. I believe that the laws are fixed and immutable and therefore that if we can discover the laws which move the earth we can predict, and to some extent control, natural events. The power of science depends not on its intellectual validity but on its practical success.

The second assumption that the human brain is capable of discovering the truth about the universe is much less firmly based. Here I use the word truth in the sense of the knowledge available to some hypothetical, omniscient and unbiased observer. Of course, this assumption is made not only by scientists but by academics of all varieties. Even though in theory they may admit the possibility, those who live by their wits cannot in practice dwell too long on the thought that their wits may be inherently incapable of uncovering the truth. The assertion that the human brain is too puny to understand the universe is one which has been made frequently in the past and is still sometimes heard today. Naturally enough it has rarely emanated from university circles: it has usually been made by anti-intellectual clergymen of one persuasion or another. It is associated more with the summer heat of the monkey-trial country of Tennessee rather than with the coolness of the dreaming spires of Oxford. Yet, oddly enough, the assertion of the weakness of man's mind is one which is by no means ridiculous and it receives some powerful support from biological science itself.

One needs to be neither particularly observant nor particularly arrogant to realise that the majority of the human race is capable of understanding the nature of the universe in only the simplest and crudest of terms. The truth about the universe is clearly beyond the comprehension of most men. Most human brains are incapable of framing appropriate questions, let alone of providing adequate answers. Even the greatest minds, the Einsteins, the Bertrand Russells, are all capable of committing idiocies, especially at the extremes of life. The great men's logic may be at fault less frequently than that of lesser mortals, but the capacity for error is still there. Einstein and the village clod may have very different mental capabilities, but they are merely at opposite ends of the same scale: they are not on different scales. At what age does the great mind become capable of discerning truth, at 10, at 20, at 30? At what age does it begin its descent into senility, at 40, at 50, at 60, at 70? At what level of intelligence does a man become capable of discerning the truth? At what level does he become incapable of discerning any of it? There are no absolutes and no rigid cut-off points. There is no evidence that the brain of any man at any time of life is incapable of error. There is no evidence that the collective intelligence of mankind is any less faulty. Truth has no meaning apart from that of being the conclusions found most satisfying to the best minds of any age. And there's the rub. Truth is what is satisfying to the mental processes of man, not what is seen to be true by our independent, hypothetical, omniscient observer. This is hardly a comforting conclusion for anyone whose livelihood depends on intellectual activity. It is one which is well understood and theoretically accepted by those members of the academic community who bother to think about it, yet for all practical purposes it is totally ignored. It is not easy for the modern aristocracy of the intelligent to admit that the chapel preacher may be right after all.

Given our present knowledge of the nature of man's evolution, it would indeed be most surprising if our mind's mechanisms were capable of discerning truth. Biologists believe that man arose by an infinitesimally slow progression from a level of organisation more primitive than that of a modern bacterium.

They believe that, just like his body, man's mind evolved by an almost infinite number of minute steps. Is the earthworm's nervous system capable of discovering truth; is the dog's, is even the chimpanzee's? Stupid questions, perhaps, yet the answer in each case must be negative. If indeed we have evolved from similar lower levels of organisation, it is difficult to see how in the course of these tens of millions of years man's brain acquired the capacity to do something quite new and, as we shall see later, quite useless: to discover the truth. If man's mind is capable of discovering truth, the reason for this is more likely to be found in the first chapter of Genesis than in *The Origin of Species*. The ideas of Darwin and his successors suggest no reasons why man should be capable of truth finding. If there is no God, if the evolutionary process operates without supreme direction, as most scientists believe, then there is no reason for believing in the validity of man's reason. On the other hand, if there is a God, if he did create man, then it is possible to believe that the ability to seek and find the truth may be one of his gifts. The intelligent thinker who rejects the concept of intervention by God, rejects also the idea that the conclusions of his own mind have any validity. The anti-intellectual preacher who thunders that God created man thereby suggests the only good reason for believing that man is capable of finding truth.

The fact of evolution therefore casts serious doubt on the reliability of man's mind. In working out the mechanism whereby evolution took place, the biologist has produced an even clearer pointer to the same conclusion. As far as we know, the most important mechanism in the evolution of living species is that of natural selection. Natural selection suggests that if any form of stress is put on a population of living creatures, those who most effectively respond to the stress will survive, while those who respond less effectively will die. The survivors pass on their successful characteristics to their descendants. Unsuccessful characteristics are eliminated and the nature of the population changes. Only success in survival matters. Virtue, wisdom, courage and justice are totally irrelevant. It is the unsuccessful who are eliminated, not those who are evil and

17

unjust. As far as we know, man's mind was shaped by natural selection. Man's reasoning power evolved not because it led to conclusions which were true in any absolute sense but because it led to conclusions which brought about successful courses of action. If over hundreds of millions of years of evolution our nervous systems have been tuned to come to conclusions which have survival value, why should we believe that these conclusions have any validity except in terms of survival? If the concept of natural selection is valid, our minds are shaped so that they come to conclusions which will lead to success in the struggle for life, not so that they draw conclusions which are wise, good and just.

A simple example may make the matter clearer. Imagine ten thousand years ago, two farming communities, Ferana and Sabara, living in the same valley. The rains fail and the crops are poor. Only enough grain remains to feed one community properly. The leaders of Ferana are what today we might call rational individuals. They believe that the failure of the rain is in no sense a result of human activity. They deny the power of the gods to control natural events as a reward or punishment for human behaviour. Sabara, by our standards, is much more primitive. Its leaders believe that the gods control the rain and that they can be angered by human behaviour. They believe that Ferana has infuriated the gods by failing to believe in their power. Heaven must be placated and the logical way to do this is to wipe Ferana from the face of the earth in a surprise attack. There is now enough grain to supply Sabara. Sabara survived because its reasoning processes led to a successful course of action. Ferana was destroyed because its reasoning processes, although apparently closer to what we would call the truth, led to a course of action totally ineffective in the struggle for survival. Truth for the leaders of Sabara is that the gods require placating: Sabara survives and its leaders are proved correct. For man truth is survival and survival is truth: error is death and death is error. History is the propaganda of the victors.

The biological view that man's mind is so constructed that it leads to conclusions which are of survival value rather than to ones which are true or just in any absolute sense thus seems to

18

be valid. Science is therefore founded on two unproven and unprovable beliefs, that the universe is orderly and that man is capable of unravelling its mysteries. Yet no scientist I know is ever racked by doubts because of this. Each gaily ignores the basic assumptions and goes on his way believing that science is successful because it leads to the truth. The truth has nothing to do with it. Science is successful because it leads to successful practical courses of action.

Chapter 3

The Nature of Scientific Research

Every schoolboy knows that science progresses by means of a logical, ordered sequence of events called the scientific method. Every schoolboy is wrong. Science is a thoroughly disorderly and illogical activity. The making of a great scientific discovery is as personal and idiosyncratic as the writing of a great poem. The major difference between the two is that with the poem it is the subjective brilliance which matters: objective validity is almost irrelevant. With a scientific discovery it is the objective validity which is vital: brilliant ideas and masterly experimentation are irrelevant if they do not lead to conclusions which are in accordance with the observation of natural events. The route by which a great discovery is reached may be personal, idiosyncratic and illogical, but the end result is not. If the end result is objectively valid, the disorderly process of discovery is justified: if the end result is invalid, the brilliance, the passion and the idiosyncrasy of the route to it count for nothing. This perhaps explains why highly individual artistic people so often dislike science. Artistic work does not stand or fall by the validity of its conclusions: the means is as important as the end. If a particular poem expressing some thought in a particular way has not been published, no rival poet, even though he may think along similar lines, is likely to publish the same poem first. There is no question of priority, because the way in which the thought is expressed matters more than the thought itself. In contrast, if a particular scientific discovery has not seen the light of day it may be made by any one of a number of men. Each may arrive at the same point by a different route, but arriving is more important than travelling hopefully. The man who gets there first is remembered, the others are forgotten.

How, then, does a scientist set about his work? First, he must find a problem which he is interested in tackling and which he thinks capable of solution. This is the crucial stage in the process of discovery. The scientist who is not interested in a problem is most unlikely to get very far with it. The one who selects an interesting topic which is insoluble with the intellectual and technical equipment available is merely going to waste his own time and other people's money. There are no prizes in this world for failing to solve important and interesting problems. That is why, to the layman's astonishment, so few people are working on such apparently fascinating topics as consciousness or extra-sensory perception. Most scientists feel that, with the techniques at present available, research in these fields is not likely to yield significant results. I have a disquieting suspicion that a similar situation may be true with cancer research. Were it not for the tremendous amount of money available, wrung from a susceptible public by an obviously emotional appeal, I suspect that very few scientists would be working directly on cancer. It seems to me that the mechanism of cancer is most unlikely to be discovered until pure biologists have unravelled a great deal more about the fundamental processes which go on inside a normal living cell. Most competent biologists probably realise this and avoid working directly on the problem of cancer. As a result, many research workers in the cancer field, I fear, are equipped with more dedication and money than scientific ability.

The successful scientist must therefore be capable of choosing problems which both interest him and are soluble. The great scientist is the one who throughout his life chooses important problems which others have found insoluble. Unlike them, he can see a way around the difficulties. This, the selection of the right problem, is the foundation of all worth-while scientific research, but it is rarely mentioned in learned discourses upon the scientific method. It is a process which depends more on intuition and inspired dreaming than on any cold, logical analysis of the situation.

Of course, even before a problem can be chosen, something must be known about the topic involved. There must be a body of knowledge, almost always confused and ill-organised, which leads to the revelation that here is a problem worth attacking. Once the nature of the problem has become apparent, a hypothesis as to the possible answer must be proposed. A hypothesis, crudely speaking, is an inspired guess which attempts to explain the observed facts in terms of cause and effect. Often, initially, the information available is too scanty to enable any sensible hypothesis to be put forward. In this case, between the recognition that there is a problem and the proposal of a hypothesis there must be an intermediate stage during which more relevant facts are collected.

In the collection of facts, the recognition that a problem exists and the formulation of a hypothesis, science does not differ significantly from many other forms of intellectual activity. In the next stage, the attempt to disprove by controlled experiment the validity of the hypothesis, science is utterly different. An experimental situation must be devised which will lead to one result if the hypothesis is valid and to another if it is invalid. Bitter experience has shown that the result of a single event like this can never be relied on. All sorts of human and technical errors may lead to a false outcome. Therefore the experiment must be such that it is capable of being repeated again and again and again. Only if the result is consistent after many trials can it be said that the hypothesis has or has not been disproved. If the idea seems valid under one set of experimental conditions it must be repeatedly tested in different ways under differing circumstances. Gradually, if the hypothesis remains undisputed, it eventually comes to acquire the status of a scientific law. On the other hand, if a reliable experimental result is at variance with the hypothesis, the latter must be discarded and the result added to the list of facts known about the phenomenon. A new hypothesis must be devised which takes into account the new as well as the old.

Thus science involves a range of different forms of intellectual activity. Few men are equally adept in all the different spheres. A scientist may be a good fact gatherer, a brilliant recogniser of problems and dreamer of hypotheses, or a clever experimenter. Only the greatest scientists are equally good at all these.

Each of the stages in tackling a problem scientifically will now be looked at in more detail.

The Recognition that a Problem Exists

There are many reasons why a man may become interested in some natural phenomenon. At one extreme is simple curiosity. Something is noticed, the behaviour of an angry bird near its nest, the colours of a film of oil on water, the way the clouds bank behind a hill before a storm, and interest is aroused. The man wants to understand the reason for the event because he is curious, not because he has to build a career, nor to make a great deal of money, nor because he wishes to help struggling humanity. Such a person is the amateur naturalist who, for curiosity's sake alone and with no thought of personal gain, takes an interest in birds, or flowers, or insects or fish. In earlier centuries perhaps the major part of scientific research was carried out by men of independent means just for the pure fun of finding out. Today, research has become so much more complex, expensive and time-consuming that only in the field of natural history can the amateur still make significant contributions.

Another quite different kind of person is yet like the amateur naturalist in that he works without thought of personal gain. A problem forces itself upon his attention, perhaps because of the human suffering involved, perhaps because of some intensely vivid personal experience. A change akin to religious conversion occurs in the man's personality. Driven by compassion and by the desire to help humanity he devotes his life to the struggle to find an answer.

But almost all important scientific research is done neither by the amateur nor by the pure and dedicated soul. Medical research, for example, is largely performed by men who certainly do not choose their problems for selfless and disinterested reasons. Some tackle a topic because they hope it will bring them money, some because it will advance their career, some because it will bring them international fame, and many – unfortunate souls – because their employer or a senior colleague tells them to do so. Yet no matter what their motive, all successful research workers have one characteristic in common, and that is curiosity. Without curiosity it is impossible to become a worth-while scientist. It is the one absolute essential.

23

The Gathering of Further Information

This is an activity which fascinates some scientists while others find it merely tedious. Some are so in love with ideas that they cannot wait to reach the stage of hypothesis formulation. They dream up hypotheses when most scientists would cautiously say that the information available was inadequate. They dream not only about their own work but also about that performed by others. For this reason their colleagues may tend to regard them as pains in the neck. No one likes to gather laboriously a great deal of factual knowledge only to have it neatly explained by someone who did none of the hard work. Yet "We are the music makers and we are the dreamers of dreams . . . we are the movers and shakers of the world for ever it seems". These infuriating purveyors of ideas are essential for scientific progress and they keep everyone else on their toes, even if only by annoying them. They irritate intensely those pedestrian men who feel that they never have enough information for the proposal of a hypothesis. Unfortunately, for the pedestrians, it is the dreamers who tend to steal most of the scientific glory. More prosaic men can hardly be blamed for grumbling when others reap what they have sown.

It is important to appreciate two requirements which information must fulfil if it is to be useful in science. These requirements influence the whole of scientific activity, they set the limits of scientific validity and they determine the direction in which science moves. First, the information must be in a form which is measurable. Second, a reliable instrument must be available for making the measurements.

An essential part of the scientific description of a phenomenon is that it must be in terms which will mean the same thing to all observers. This means that words are virtually useless for, like Paul, they are all things to all men. Only by making numerical measurements can any reliable degree of objectivity be introduced. Only when in the form of measurements can the observations of two human beings be reliably compared. Only when in the form of measurements can the observations made when a scientist is young be objectively related to those made when he is old. Only measurements retain their integrity from observer to observer. Ten centimetres is ten centimetres whether I am boy or girl, Russian or American, Arab or Israeli. Helmholtz even went so far as to say "Science *is* measurement". Coupled with its corollary, "Science is objective comparison", this makes not a bad description of what scientists spend their lives doing. In this description lies the key to both the strength and the weakness of science. In theory the conclusions of science do not depend on who is making the observations. They depend on comparison with the supreme arbiter, nature, whose laws we believe do not change. Scientific conclusions stand firmly on the foundation of a universal, unalterable standard. In this sense they are far more reliable than are conclusions not based on measurement. On the other hand, the weakness of science is that it cannot be applied to qualities which do not lend themselves to objective numerical assessment. And it is precisely these qualities which most human beings find interesting. How does one measure beauty, courage, strength or weakness? It certainly cannot be done scientifically. These qualities cannot be reduced to centimetres and kilogrammes. These most interesting of topics lie for ever outside the realm of science.

Consider a real example of the importance of measurement. Twenty years ago a physician involved in the treatment of a patient with cancer of the lung might have said to a colleague, "You know, I get the impression that an awful lot of these chaps with lung cancer are heavy smokers". This is an important observation, the first glimmering of an idea which may be worth investigating. Unless carried further it is scientifically useless. Impressions are notoriously unreliable unless they are confirmed by measurement. Like old men remembering lost summers when the sun always shone and the rain fell only at night, doctors who have an idea remember only those patients who confirm it and forget those who do not fit in. The next stage, therefore, is to ask all patients with cancer of the lung about their smoking habits. The results are then scientifically valuable. They pinpoint the incidence of cigarette smoking in lung cancer sufferers in that particular hospital and at that particular time. The vague impression has been reduced to reliable figures which can be compared with the incidence of smoking in normal people, with that in lung cancer patients in other parts of the country and of the world, and with that in the same hospital in succeeding years. Gradually, in this way, there can be built up the picture we have today. We know that among those who do not smoke, lung cancer is a very rare disease. We know that even five cigarettes per day increases the risk and that heavy smoking of over twenty per day amounts to the reckless exposure of oneself to a killer. Thus, from studying the behaviour of ordinary people, important scientific conclusions have been drawn. The medical profession is now faced with the problem of making the scientific conclusion such a part of popular folklore that the incidence of cigarette smoking and of lung cancer become greatly reduced. Unfortunately, all the signs are that success in this endeavour is elusive.

The essential condition that a subject studied must in theory be amenable to objective measurement greatly limits the scope of science. Its range is narrowed even further by the practical need for an instrument capable of making the measurement accurately. Many phenomena which in theory are capable of objective assessment cannot be studied in practice because we do not yet have the methods available. An inaccurate measurement is worse than no measurement at all. The latter cannot mislead, the former can. The lack of a satisfactory method may prove such a stumbling block that a great deal of tedious effort must be put into the design of a suitable instrument. On the other hand, the invention of a new technique may open up a hitherto completely unknown field and lead to the development of whole new sciences. It is at this stage that many problems must be abandoned because no adequate instruments exist. It is also at this stage that many scientific reputations are made, when some inspired individual sees a way of making a measurement which has hitherto been technically impossible.

Medical science again provides an excellent example. The heart is a pump which raises the pressure of the blood passing through it. In this way the blood can be forced along the arteries to every organ in the body. Pressures which are too high and too low are both dangerous and there are many situations in which doctors wish to be able to measure the arterial pressure. Until the last quarter of the nineteenth century this was virtually impossible. Methods of measurement of fluid pressure required direct access to that fluid. In man this meant sticking a needle with a pressure sensing device directly into an artery. This is painful and technically not easy, and as a routine method it was and is out of the question. Doctors had to guess the blood pressure by feeling the pulse, and this is a thoroughly unreliable method. Then, in a brilliant flash, Riva Rocci saw how human arterial pressure measurement could be made much more accurate. He realised that the peak of arterial pressure reached when the heart is contracting (systolic pressure) could be measured quite easily, without pain and with the minimum of discomfort. A long flat rubber bag is attached to a device for measuring the pressure within the bag. The bag is wrapped around the upper arm and blown up until no arterial pulsation can be felt at the wrist. By means of a valve, the pressure in the bag is then slowly reduced until the pulse just reappears. At this point the pressure in the artery when the heart contracts must be just able to overcome the pressure in the bag. A spurt of blood bursts past the tourniquet and the pulsation can be felt at the wrist. The pressure in the bag at this point therefore gives an approximate measure of the systolic pressure.

Much modern medicine depends on this simple method. There are few diseases in which an accurate knowledge of the blood pressure may not be important. The pressure may be dangerously lowered by haemorrhage or by coronary thrombosis. Other organs may then receive an inadequate supply of blood and be themselves in danger of damage. On the other hand, the blood pressure may be much too high, as in some types of kidney disease. There is then a risk of small vessels becoming damaged and bursting under the strain. In particular the kidneys, the eyes and the brain are at risk. Most people who suffer a stroke, for example, have too high an arterial pressure. We now have drugs both to lower and to raise the blood pressure, but if they are to be used safely it is obvious that the dosage must be controlled by actual pressure measurements. This can be simply done, thanks to Riva Rocci's brilliant yet elementary concept. Doctors now use a stethoscope to listen for the spurts of blood overcoming the resistance of the rubber cuff because this has proved to be more accurate than feeling the pulse. But the principle is the same as the one first proposed nearly a hundred years ago.

The Formulation of a Hypothesis

Many find this the most exciting and stimulating part of science. This is the point at which individuality and creativity are most important. A hypothesis is never formulated by a team. The germ of the idea occurs in the mind of a man alone. Measurements may be made, experiments may be carried out and hypotheses may be modified by teams, but the original idea almost always depends upon the dreaming of one mind. It is at this point that science comes closest to the arts.

It is not sufficiently appreciated that hypotheses come in two quite different forms. I shall call these the major and minor varieties. The minor hypothesis is a relatively low form of intellectual activity of which the majority of the human race is capable. It depends on an understanding of the simpler rules of logic and of the principle of cause and effect. A million garage hands and home handymen formulate minor hypotheses every day as they endeavour to put into working order some piece of machinery which has failed. Doctors, too, carry out the same procedure as they attempt to diagnose the nature of a disease.

The essential features of the minor hypothesis are:

1. A problem presents itself which is similar to other problems which have occurred many times in the past. A car will not start, an electric light fails, a man walks into a doctor's consulting room with a pain in his stomach. In each case one wants to know the cause of the defect and in each case the number of possible correct answers is limited.
2. A few simple tests are carried out. On the basis of the results

a hypothesis is formulated as to the nature of the trouble and the best way of tackling it.
3. Practical measures are applied to correct the defect. If the car starts, the light shines or the man loses his pain, the hypothesis was probably correct. If the situation remains unchanged, the original hypothesis must be rejected and a new one devised.

Most scientific activity is conducted on this level. Well-tried logical methods are used to solve problems which differ only in minor ways from similar problems which have been successfully tackled previously. Designing a new lipstick or a new refrigerator, devising a modification of medical treatment, constructing a bridge or building a dam are all problems for which reliable methods are available. One merely has to apply logically a few principles which have been hammered out in the past. No creative genius is required. The examples are all taken from applied science or technology, science devoted to solid practical ends. They have been chosen because lipsticks, doctors, refrigerators, bridges and dams are all concepts with which everyone is familiar. It is often assumed that so-called pure science (science not devoted to any immediate practical end) is a different order of activity requiring a much higher level of mental power. This is just not true. Physics, chemistry, biochemistry and biology all largely involve the application of old and well-used technical methods to problems which are little different from problems which have been dealt with before. The minimum of thought is required. All that is necessary is to follow with suitable trivial modifications the proven recipes developed by previous generations of chefs. Both our universities where pure science is king and our industries dominated by technology are largely staffed by prosaic individuals tackling prosaic problems with what amounts to not much more than complex kitchen equipment and cookery books. As far as both the practical techniques and the thought processes are concerned, there is absolutely no difference between science pure and science applied. There is no justification whatsoever for the almost moral superiority which some affected academics claim over engineers and technologists.

The development of a major hypothesis requires a quite different order of intellectual activity. Major hypotheses are rare events. Scientists capable of the necessary conceptual leaps are few in number. In retrospect the major hypothesis is always seen to be beautifully logical. But in contrast to the minor hypothesis, conscious logical thought processes play no part in its discovery. What I mean can best be illustrated by actual examples. The fall of Newton's apple is perhaps the most famous scientific event ever recorded. The fact that objects fall to earth was known to all mankind. Newton himself had observed this uncountable times before. Looking back it seems completely logical to suggest that this behaviour can be accounted for if a force exists which attracts all bodies to one another. Yet this had occurred to no one until Newton, dreaming in the orchard, saw the apple fall. This was the spark which kindled the flame of insight in Newton's brain. In an instant he saw that a force of gravity could account not only for the falling of objects to earth but also for the behaviour of the moon, the planets and the sun. Once seen, it all seemed perfectly logical, and it only remained to measure and define the magnitude of the force. The behaviour of falling bodies, of the sun, the moon, the earth and the planets had fascinated mankind for many years. Yet all the careful thought, all the wrestling with the problem, came to nothing until the simple falling of an apple provided the stimulus which put everything in place.

Another example is that of Kekule and his discovery of the chemical nature of many organic compounds (chemicals associated with living things). By the middle of the nineteenth century it had been established that the universe is built up from minute particles known as atoms. There are about a hundred different sorts of atoms known as the elements. Most natural substances are not pure elements made up of atoms: they are compounds made up of other tiny particles known as molecules. Each molecule consists of several atoms bound together, and the number of atoms of each type in any given molecule determines the nature of that molecule. Water, for example, consists of molecules, each of which contains two atoms of hydrogen and one of oxygen. If two atoms of oxygen are combined with two atoms of hydrogen, the quite different molecule of hydrogen peroxide (hair bleach) is the result. In Kekule's time, organic materials seemed to be made up of molecules which contained very large numbers of carbon atoms. No one could understand how such large numbers of carbon atoms could be fitted together to form a stable structure. It was suggested that they were linked in long and complicated chains, but this seemed a very cumbersome way of doing things. Kekule, like many others, had been working on this problem without any very satisfactory result. One day he sat dreaming by his fireside in a semi-conscious state. Snakes of various lengths and colours danced mysteriously before his eyes. Suddenly one of these, instead of writhing along in a normal snake-like manner, took its own tail into its mouth. At once Kekule saw the answer to the central problem of organic chemistry. Instead of being linked in long, unwieldy chains, the carbon atoms must be arranged in rings. Looking back it is almost incredible that no one had suggested it before. Kekule's hypothesis proved correct and we now know that the ring of six carbon atoms is the unit on which many organic compounds are based. The answer is perfectly logical, yet no conscious, formal logical processes were involved in its discovery.

The major hypothesis therefore presents a radically different answer to an old problem. It follows no established rules. It redirects the stream of scientific endeavour. The two examples were taken from pure science, but the major hypothesis is just as important in the applied field. In both types of science, two varieties of major hypothesis can be found. They are both well illustrated by the story of *The Dam Busters*. During the second world war, Germany's military might depended to a large extent on the industrial power of the Ruhr. In turn the Ruhr industries depended on the water and power provided by three massive dams, the Mohne, the Eder and the Sorpe. Obviously their destruction could be a great prize. The British Royal Air Force studied the situation carefully and reluctantly came to the conclusion that bombing would be ineffective. The concrete was so thick that in order to burst the dam a huge explosive charge would have to be laid deep in the water right against the wall itself. If the charge exploded away from the concrete, much of the force would be dissipated by the water. With conventional bombs dropped from a moderate height the necessary accuracy would be quite impossible to achieve. The bomb would have to hit the water at precisely the right angle and at precisely the right point within a yard or so. Even if this miracle were achieved, the bomb would almost certainly be deflected during its passage through the water. In almost every case it would explode well away from the vital point.

But Dr. Barnes Wallis (now Sir Barnes Wallis) brought up a seemingly ludicrous idea. He suggested that low-flying aircraft at a precise height above the water should drop a new type of bomb far from the dam. Like a skimming stone it would bounce along the surface, gradually losing momentum until it plopped into the water just behind the wall, sank straight down and exploded in precisely the right place. Wallis had absolute faith in his idea. Most of the other experts were sceptical and most of the early tests failed. The bombs either sank or broke up. Yet Wallis still believed and fought a bitter battle to gain support. At last he produced a bomb which really did work. The Royal Air Force went ahead with the raid and two of the supposedly indestructible dams were destroyed.

This is an excellent example of a brilliant dream, thought crazy by all but the dreamer, and proved valid only after a prolonged struggle. The answer seems just too outlandish to be true. The experts themselves have so little faith that they are even more opposed to the idea than are the laymen who must eventually say yea or nay to the whole project. Yet the experts are wrong. If only there were a way of giving reliable guidance as to what crazy scheme should be adopted and what should not, the relation between science and government would be much easier. It is regrettably true that the great majority of scientists' dreams are worthless. As one man in a position of great power said: "Of all the inventors' schemes put before me, 99 out of every 100 are useless. Unfortunately there is no certain way of picking the good from the bad. If I back any ideas at all I am almost certain to be wrong more times than I am right. If I reject every one I shall be correct in 99 per cent of cases and what administrator could claim a better record than that?" Very sad but very true. Perhaps the only answer is to abandon the whole concept of supporting projects and to back men instead. More will be said about this in Chapter 8.

The other form of major hypothesis has a very different passage. It is so clear and brilliant that it is accepted at once without argument. Again the Dam Busters provide an example. Once the idea of the bouncing bomb had been officially accepted, the Royal Air Force had to put the project into operation. The bomb would work only if dropped from a low-flying aircraft at a fixed height above the water. No altimeter available could measure height with the required accuracy. All were hopelessly unreliable at that low level. As the night for the attack approached everyone became more and more concerned about the possibility that the whole magnificent project might founder on this one obstacle. One night a group of officers was at a cabaret. They watched, fascinated, as a girl danced, caught in the twin spotlights. One man suddenly saw the answer. Two spotlights could be fixed at opposite ends of the aircraft at such an angle that their beams would cross the required distance below the aircraft. When the two spots appeared as one, the height of the aeroplane must be correct. The idea needed only to be explained for it to be instantly accepted. It was so startlingly clear that there was no need to work hard to convince people of its validity.

Concepts in pure science are also sometimes instantly accepted in this way. One example is the structure of DNA (deoxyribonucleic acid) proposed by Watson and Crick in 1953. DNA had for several years been recognised as being a most important constitutent of living organisms. It seemed to be the chemical responsible for inheritance and also for the normal everyday control of cellular activity. Several great scientists recognised that the discovery of DNA's structure would be a great prize and they were determined to win it. But Watson and Crick, two young unknowns working in Cambridge, England, were the first to get there. They saw that all the properties of

DNA could be accounted for if two substituent strands were twisted around one another in a helical formation. The story has been vividly told in James Watson's book *The Double Helix*.[3] Incidentally, scientists were right in their estimation of the worth of the prize. From the Watson–Crick structure has come a great stream of brilliant discoveries. Even material reward was not lacking. The discoverers received the Nobel Prize in 1958.

I believe that the formulation of hypotheses is greatly affected by the way in which scientific research is organised. In recent years the tendency has been to employ larger and larger teams to tackle a problem. The lone worker has gone out of fashion. I am doubtful as to whether this will lead to more rapid scientific progress. If progress is assessed in terms of the accumulation of fact or in terms of the number of scientific papers published per year, there can be no doubt that teams mean rapid advance. But if progress is assessed in terms of the development of important radically new concepts then I am not so sure. The team is a highly effective instrument for rapidly covering ground which in principle has been covered before. It is extremely successful in the areas which can be satisfactorily dealt with by minor hypotheses. But I do not believe that teams ever produce the startling ideas which change the course of science. Such hypotheses are much more likely to be born in the mind of one man or, as in the case of Watson and Crick, by two open and brilliant intellects spurring one another on. The team is just as likely to suppress major conceptual advances as it is to make them. All teams consist mainly of individuals incapable of stepping outside an established line of thought. Many major hypotheses seem ludicrous at first sight. If the decision to investigate the idea or not depends on the team, the conservatives are likely to win. A man alone can follow his crazy hallucinations without having to drag the rest of the team with him.

Of course, if one wants to be certain of a safe, solid and distinguished career in science, then joining a team is the way. One makes contacts who will help one along. One is to some extent insured against making a fool of oneself. Even if this happens it is in the company of four or five others. One gets far more publications to one's name. Six men working together can pour out at least six times as many papers as one man working by himself. The member of the team gets his name on all the papers produced by it and so has a life list of work several times longer than that of the loner. In days when appointments committees do not actually read the candidate's work but merely look at the length of the list, this is an important consideration. The loner's life is much less secure. He makes fewer contacts, he is more likely to show himself a fool, his list of publications is short. And yet he is a free man. He can follow to the end his own star. If his name stands alone, no one can whisper that it was really old so-and-so who did all the work. In science, as in most of life, "Down to Gehenna or up to the throne, he travels the fastest who travels alone".

The Characteristics of a Good Hypothesis

A good hypothesis has three major characteristics. It accounts for most of the facts available. It accounts for those facts in a precise, direct way. It makes predictions which are amenable to experimental testing and which suggest the direction in which further progress may be made.

There is no need for an hypothesis to account for *all* the facts available. The history of science is littered with so-called facts which were later found not to be facts at all. Pupils in school are taught all about the supremacy of scientific fact and about the necessity of subordinating hypothesis to it. In reality both fact and hypothesis are almost equally fallible. Anyone who has ever worked in a laboratory, particularly a biological one, is fully aware of the vulnerability of experimental fact. Experiments are always going wrong and must be repeated again and again until the "right" result appears. It is often extraordinarily difficult to persuade two similar experiments to give precisely the same result. Ill-understood factors are always intervening and making a mess of things.

A famous example of the conflict between hypothesis and experiment concerns the brilliant physicist and Nobel prize winner, Schrodinger. In 1926 he developed a mathematical hypothesis which seemed to describe beautifully the behaviour of the electron, one of the sub-atomic particles of which atoms are made. Unfortunately, this did not fit in with some of the known experimental facts, and Schrodinger modified his equations. The concept then did not seem so beautiful and satisfying, and publication of the work was delayed. A short time later it was found that some of the experimental facts were not facts at all. Further experimentation demonstrated that Schrodinger's original unmodified work was in accordance with reality. If he had had more faith in his hypothesis he would have ignored the facts and published his work anyway. Of this episode another Nobel prize winner, Paul Dirac, has said: "It is more important to have beauty in one's equations than to have them fit experiment."[4] In other words, it is more important to have a beautiful and satisfying hypothesis than to have one which fits all the so-called facts. The man who has faith in his dreams must be prepared to fight for them in the face of experiment. If he is to make great discoveries he must be arrogant enough to believe that he is right and the experimental facts are wrong. On the other hand, he must be humble enough to remember T. H. Huxley's aphorism: "The tragedy of science is a beautiful hypothesis slain by an ugly fact." Amid all his arrogance the great scientist must not be so drunk with dreaming that he forgets that beautiful hypotheses can be wrong, that they can be destroyed by reliable experimental work and that they can be replaced by yet more beautiful dreams. There is nothing more inspiring than the story of a Barnes Wallis who fights on against all the evidence and is eventually vindicated. There is nothing more pathetic than the story of a man who continues to believe, in the face of all the evidence, and who in the end is forced to admit that he was wrong. In science it is the end which matters, and that is both its tragedy and its glory.

A good hypothesis must not only account for experimental facts. It must do so in a precise, direct way. The connection between hypothesis and fact must not be too vague. In addition, predictions must be made which allow the hypothesis to be tested.

Again the connection between hypothesis and prediction must be close. It is clearly desirable that a hypothesis should be correct, but mere correctness without precision and testability is useless.

To take a simple example, suppose that your car has not been serviced for several years and that in spite of this it has given you a long period of trouble-free running. It suddenly breaks down at an awkward moment and you telephone for a mechanic. On arrival he takes a high moral line and tells you that the reason for your breakdown is lack of servicing. He suggests that he should tow your car to his garage and give it a thorough 5,000 mile check. The hypothesis is almost certainly correct, but for you who are in a hurry it is useless. It is too imprecise to pinpoint what is wrong and it fails to make any predictions as to how the matter may be rapidly put right. In disgust you telephone another garage and a second mechanic arrives. His first suggestion is that the sparking plugs are faulty. This hypothesis, unlike the first one, may well be wrong but it is much more useful. It is precise, it can be tested, and it predicts that changing the plugs will make the car go. This illustrates an important point which is not understood even by many scientists. A precise, testable hypothesis is useful even if it is wrong. Suppose that the test demonstrates that the sparking plugs, despite their neglect, are working satisfactorily. This enables the sparking plugs to be eliminated as the cause of the trouble and the search can be directed to other parts of the mechanism. Further precise hypotheses can be proposed, tested and rejected if necessary until the fault is found. The importance of such precision can never be overemphasised. Knowledge is advanced if such hypotheses are wrong as well as if they are right. A precise hypothesis can be rapidly tested and if it proves to be faulty the area it deals with can be safely ignored. If it turns out to be correct, further investigation can be carried out along the same lines. If no hypothesis is proposed and if information is collected in a vague, woolly sort of way, progress is certain to be slow and erratic because it is unstimulated by precise questioning. It is often a long time before the profitable and useless fields of investigation become satisfactorily delineated. The mechanic who suggests that a tow to his garage, followed by a full service, is the answer to your trouble will undoubtedly get the car going again eventually. But if you want to get to your destination quickly, the second mechanic will be much more useful.

The Testing of a Hypothesis

Science differs significantly from other forms of intellectual and cultural activity at only one point. The scientist must always be prepared to test his creative ideas by referring the argument back to nature. In the arts, success is measured by human response, by the degree of critical acclaim, by the size of the circulation which a book achieves or by the amount of money people are prepared to pay for one's painting. In science all these things are largely irrelevant. Success depends on comparison with an inhuman, universal standard. Herein lies both the power and the irrelevance of the appeal of science to the human mind.

The artist in any medium can in the last analysis never be proved wrong. No matter how many disagree with his ideas, he can always aver that their conclusions are matters of opinion. Critics may universally condemn his work, but the artist can legitimately reply that their judgment cannot be conclusive. In this he has history on his side. There are innumerable examples of artists ruined by contemporary criticism and vindicated many years later. Equally the artist who is acclaimed may later find the wheel turning against him. No artist can be proved wrong, nor can he with any certainty be proved right. He cannot be reliably justified or condemned by an external authority.

The scientist must always be prepared to submit to such authority. He can be proved wrong by objective arguments. He cannot maintain that the inadequacy of his work is merely a matter of opinion. This restraint on the individuality of the human mind is the negative side of science. There is a positive side. The scientist can have the satisfaction of being proved right by reference to nature. He can know that his work is valid in a way in which the artist never can.

Hypotheses are referred back to nature by means of the process known as the scientific experiment. This is by no means such a cut and dried procedure as one might imagine from accounts in elementary scientific texts. A hypothesis is essentially a statement of cause and effect. It states that x occurs as a consequence of a, b and c and so on. An experiment is an attempt to test the particular cause and effect sequence suggested by the hypothesis. The principle involved may be illustrated by a concrete example from my own work.

One of the most important unsolved problems in obstetrics today is that of toxaemia of pregnancy (pre-eclampsia). This is a common condition which tends to develop in the last two months before the child is due. The mother gains weight very rapidly. The blood pressure is raised. Damage to the kidneys leads to the appearance of protein in the urine, an abnormal event. Fluid is not properly eliminated from the body, leading to a puffy face and swollen ankles. Rarely toxaemia can cause fits and, in the worst cases, death of the mother. But the danger to the mother is negligible compared to the risk to the unborn child, and herein lies the real importance of the disease. Babies born to toxaemic mothers run much higher risks of dying either in the womb or soon after birth than do babies born to normal mothers. Even in those who survive there is an increased incidence of mental retardation. When I was a medical student I saw a forty-year-old woman who had married late and who desperately wanted a child lose her baby because of toxaemia. This aroused my interest in the problem. Of course, I was not uninfluenced by the fact that relatively few good scientists were working in the field and that because of its importance anyone who answered the problem would reap great rewards in the scientific heaven.

First, I had to have a hypothesis. I knew that towards the end of pregnancy two chemicals are produced in large quantities by the placenta. These are oestrogen and progesterone. I also knew that in animals the injection of progesterone stimulates appetite and causes excessive weight gain. My first tentative hypothesis was that progesterone was the villain. I suggested that for some unknown reason progesterone levels in toxaemia were unusually high and that this high progesterone level was responsible for the weight gain, for the high blood pressure, for the damage to the kidneys and for the retention of fluid in the body. I mentioned this idea to some obstetrician friends. They all pointed out that it was untenable because progesterone levels in toxaemia are lower than normal. Study of several modern textbooks confirmed that they were right. However, by this time I had fallen in love with my hypothesis and I was not satisfied with the validity of this fact with which all obstetricians seemed conversant. I went to a library and looked up the original research publications upon which the statements about the low progesterone level in toxaemia were based. To my delight I found a flaw in the argument. The widely accepted fact that progesterone levels in the body in toxaemia were low was not based on measurements of blood progesterone. Apparently it is very difficult to measure the amount of progesterone in a blood sample and the results of the measurement are unreliable. But the active substance progesterone is destroyed in the liver and converted to an inactive chemical known as pregnanediol which is excreted in the urine. Urinary pregnanediol levels are relatively easy to measure. It was argued, not unreasonably, that the amount of pregnanediol appearing in the urine should be roughly proportional to the amount of progesterone in the blood. The output of pregnanediol in toxaemia is low. It was therefore assumed that progesterone levels in toxaemia must also be low. Again the argument is not unreasonable. It was on this basis that the obstetricians thought my ideas untenable.

But there is one loophole. Suppose that the liver is in some way at fault in toxaemia. Suppose that it fails to destroy progesterone normally. The amount of pregnanediol formed will be less than normal and the amount excreted in the urine will be small. At the same time, because progesterone destruction is ineffective there will be higher levels in the blood than usual. A low urinary pregnanediol output might actually mean that blood progesterone was high rather than low.

This meant that my hypothesis could still be correct. The most certain way to check it directly would be to measure blood progesterone and urinary pregnanediol levels in normal and toxaemic women at different stages of pregnancy. Unfortunately, at present with the resources I have available, this is not a practical proposition. Methods for measuring blood progesterone are still cumbersome and of uncertain accuracy. I know what I want to measure but I cannot measure it. This illustrates clearly a common scientific predicament. Research is often held up not because of a lack of ideas but because of a lack of techniques to make the desired measurements.

But there are other experiments which may support or disprove my hypothesis. Animal experiments have demonstrated that progesterone injections can cause weight gain, but until recently no one had demonstrated that such injections could raise blood pressure. Therefore I decided to investigate the effect upon blood pressure in rabbits of giving progesterone injections. With high hopes I obtained some animals and an apparatus for measuring blood pressure and began work. After a few days, to my intense disappointment, the experiment did not seem to be turning out as I had hoped. Instead of rising, the blood pressure actually fell. This time my hypothesis really did appear to be invalid. However, I had no other bright ideas with which to replace it and just for the sake of interest I decided to keep on with the injections for a few more days. In two rabbits the pressure remained low, but in another six it returned to normal and then rose well above normal. On giving dummy injections without progesterone, the pressure fell. On restarting progesterone, after a few days pressure rose again. Thus largely by chance I have discovered another piece of experimental evidence in favour of my idea. I do not know how the progesterone works and I do not yet know whether it will have the same effects on human beings. However, the contraceptive pill contains progesterone-like compounds and there have been several recent reports of women on it developing high blood pressure. This returned to normal when the pill was stopped. The unknown mechanism may therefore also operate in human beings.

The story so far, simplified and incomplete as it is, gives some idea of how science operates in practice rather than in theory.

41

Twice my hypothesis appeared to be invalidated by experimental facts. The first time when the experts told me that it was a well-known fact that progesterone levels in toxaemia were low. Only because I believed in my ideas, without any very solid evidence to support my beliefs, did I bother to investigate the matter carefully. Only then did I find that the supposed fact was not a fact at all. It was a deduction from an experimental finding. The logical link between experiment and deduction seemed reasonable enough but it contained one dangerous flaw.

The second occasion was the finding that progesterone injections actually lowered blood pressure in rabbits. This was a genuine experimental result. Only by chance did I continue the experiment. I had no reason to believe that the long and short-term results of the injections would be different. If I had stopped the work after a few days, the half truth that progesterone lowers blood pressure might have passed into the literature. It might have been many years before someone chose to ignore this experimental fact and investigated the matter more carefully.

Another point that becomes apparent is that it is essential to repeat an experiment many times before one can be sure that the results are valid. Most of the effects studied in science can be brought about by many causes. This is particularly true of blood pressure. Especially in the early stages of the study of a phenomenon, many of the factors which influence that phenomenon may be unknown. If they are unknown, they cannot be controlled by the experimenter. Therefore, as he tries out his postulated cause and effect sequence (in this case the action of progesterone on blood pressure) the results may be influenced by factors of which he is quite unaware. If this happens the experiment may be invalidated, but the experimenter may not be aware of this. If my experiments had been done on two rabbits only, by chance I might have picked on the two animals in which the blood pressure failed to rise. The lack of any effect of progesterone in raising pressure would have become an experimental "fact". I do not yet understand why some animals respond to the chemical and some do not. If I can discover the reason it may provide a clue as to why some women develop toxaemia and some do not. An enormous amount of experimental work remains to be done before I can be reasonably sure that my hypothesis is correct. So far things have gone well. I am a believer myself, but it will be a long time before I have many converts.

In summary, the experimental approach to nature is the only point at which science differs radically from all other approaches. There are four essential features:

1. It is essential to have a precise hypothesis which is capable of being tested. Untestable hypotheses are scientifically useless.
2. Well tried and accurate methods must be available for measuring the factors under study.
3. The experimental tests of the hypothesis must be such that they can be repeated over and over again. The results of a single piece of work can never be relied upon. Only rarely do we know all the factors which influence a phenomenon. Only rarely can we be sure that some unknown factor is not disturbing the results.
4. The scientist who formulates a hypothesis must have so much faith in it that at least in the first instance he is prepared to ignore apparently contradictory experimental facts. Most major advances in science depend on demonstrating that some widely accepted fact is either just not true or is based upon a misconception.

The Importance of Controls

One aspect of the scientific approach which is not understood by most laymen and some scientists is the importance of control experiments. The principle can again be illustrated by the experiments in which I gave progesterone to rabbits. Progesterone is supplied by the drug company dissolved in an oily substance called ethyl oleate. This is believed to be inert and without effect on any organ. But before my work no one had investigated the effect of ethyl oleate on blood pressure in rabbits. How, then, could I be sure that the observed changes in blood pressure were not due to the effect of ethyl oleate? How could I be sure that the changes were not simply caused by pricking the animal every day or even by the ageing of the rabbit? I had to do control experiments in which I did everything but give the progesterone. To some animals, therefore, I gave daily injections of plain ethyl oleate. The blood pressure remained steady. Ageing, ethyl oleate and the daily prick could therefore not be responsible for the changes. I could therefore be sure that the rise in blood pressure really was due to progesterone.

Nowhere is the proper use of controls more important than in medicine. Suppose I want to investigate the effect of a new treatment on the common cold. The simple-minded approach applied so convincingly by some manufacturers of patent medicines is to say 1,000 people were given our new wonder drug X. They all got better quickly. This proves how effective our drug is. But most people with colds get better quickly and this finding is meaningless. What we must do is to compare 1,000 people treated with X with 1,000 people treated with tablets which look just like X but which do not contain any of the active substance. Only then can we truly see whether X makes any difference. Only when proper controls are applied in this way can we be certain that the outcome of any scientific experiment is really due to the variable which we think that we are altering. Experiments done without controls are not worth the paper the results are written on.

One example of the value of a control group arose during the demonstration of the relationship between smoking and cancer. As most people now know, once this problem began to be studied seriously, statistical methods soon demonstrated a very close relationship between the smoking of cigarettes and the development of cancer of the lung. The use of statistics is most important in scientific research. It can be employed to clarify the meaning of experimental results and can also demonstrate associations between phenomena as in this case. But such statistically demonstrated associations, no matter how convincing they may appear, can never be accepted as proof that the one phenomenon causes the other. For example, it would no doubt be possible to demonstrate statistically an extremely close association between the ownership of refrigerators and the ownership of washing machines. It could be shown that almost invariably families who own a washing machine also own a refrigerator. But that does not mean that the family owns a refrigerator *because* it owns a washing machine or vice versa. The two phenomena have a common cause in that families who have a certain level of income feel that both appliances are necessary for the running of their home. Neither one is the cause of the other.

Similarly, even though lung cancer rarely occurs in non-smokers, the mere statistical demonstration of this association is not sufficient to prove that smoking causes cancer. For instance, it is legitimate to suggest that individuals with a constitution which makes them smokers also develop lung cancer as a consequence of that constitution. The smoking and the lung cancer might thus both be consequences of a common factor and the smoking might not cause the cancer. The only way to demonstrate that the causal association is real is to compare two groups, both susceptible to the smoking habit, but one of which does not smoke. Recently this comparison became possible. Many heavily smoking doctors were so impressed by the first reports on lung cancer that they abandoned the habit. Here, therefore, was a readily identifiable group of people, undoubtedly susceptible to the lure of tobacco, yet who were not smoking.

These men have been watched over the past few years and their medical history compared with that of a group of individuals of similar age and social status who continued smoking. Initially there was little difference between the smokers and the control group who had abandoned smoking. But as the years have passed, a clear distinction has emerged. The incidence of cancer is considerably lower in the group which is not smoking. The proof that smoking causes lung cancer is now, therefore, becoming truly scientific and rests on more than the mere demonstration of a statistical association. When two apparently identical groups are compared, both known to be susceptible to the smoking habit, the one which smokes has a much higher death rate from lung cancer than the control group which does not smoke. This is a properly scientific conclusion. But it is a pity that it is not more often remembered that statistical associations between phenomena do not necessarily mean that the one is causing the other. Often the association is either fortuitous or the two phenomena both depend on some common factor. A cause and effect sequence can be demonstrated conclusively only by means of a controlled experiment.

Chapter 4

Scientific Law and the Practice of Science

A hypothesis is a preliminary attempt to explain a natural phenomenon in terms of cause and effect. It must be tested by experiment and, if the results are not in conflict with it, the hypothesis may attain the level of a theory. If repeated testing over many years and under many different conditions confirms the validity of the theory, then the theory may acquire the status of a law. There is no precise dividing line between a hypothesis and a theory, and a theory and a law. Long-winded attempts by philosophers and others to produce precise and limiting definitions of these words ignore their loose practical usage, add nothing to knowledge and merely waste time and intellect.

The expressions "Scientific research proves that..." or "Such and such is a scientific law" are often thrown into arguments as though the talisman words "scientific", "research" and "law" have only to be used and the opposition will be routed. Scientific research proves that Fluz washes better than Buz, that your garden will flourish when you apply Green Fingers liberally three times a day before meals, that children not cuddled enough by their mothers want to marry their grandparents. It should now be apparent that because of the limitations of experimental reliability most such claims are nothing but incanted mumbo jumbo. Dignified professors who pronounce that such and such is a scientific law are often little more trustworthy. Belief in the absolute reliability of scientific law is so common and dangerous a misconception that the nature of such law is worth examining a little further.

A scientific law is a widely accepted generalisation which explains some phenomenon in terms of cause and effect. First it defines what factors influence the phenomenon under study. These factors are often called variables. It then defines the

47

relationship between the variables and the phenomenon. For example, suppose I am studying the amount of space occupied (the volume) by a given amount of gas in a cylinder fitted with a piston. I know that if I press the piston inwards, the volume of the gas decreases while its pressure increases. Therefore pressure is one of the variables which must be considered. I also know that heating the gas and raising its temperature causes the gas to expand and pushes the piston out. Temperature is thus another variable and, in this simple situation, temperature and pressure are the only ones which need to be taken into account. In order to work out the nature of the laws governing the interaction of pressure, temperature and volume, it is necessary to perform an experiment. The effect of altering one variable at a time must be studied. For example, I might investigate the effect of altering the pressure while the temperature remained constant, or I might study the effect of altering the temperature while I kept the pressure constant. Attempts to alter both factors simultaneously produce a complex situation whose results cannot easily be interpreted. It is difficult to be sure just how much change in volume is due to a change in temperature and how much to a change in pressure unless the two effects are first studied separately. If it is difficult in this situation, which is, in fact, extraordinarily simple, how much more of a problem is it in biological work where there are dozens of variables to be considered. It is a cardinal principle of successful experimentation that the effect of moving only one variable at a time must be studied.

As was discussed in the last chapter, no scientific law can ever be proved in a logically watertight way to be universally true. It is quite impossible to know whether a law is a valid description of events at all times and in all places. All one can say is that in all the circumstances in which the law has been tested, the description has been found to be satisfactory. There are excellent examples in physics of laws which are satisfactory descriptions of events on an earthly scale but which fail on the cosmic level. Newton's laws, for example, are good accounts of events on earth but fail on the scale of the universe. There, Einstein's theory of relativity is required. Incidentally, it is not widely enough appreciated that this famous piece of scientific work has never been experimentally proved. In fact, there is some experimental evidence which contradicts the theory, but Einstein's ideas are so attractive that few physicists can bring themselves to reject them. It is the beauty of the concept which makes it valid for them, not the experimental proof.

The scientific description of natural events reaches law status only when all the factors which influence a phenomenon are known, only when they are all measurable, only when they can be studied experimentally and only when the experimental work is consistent and repeatable. Scientific description cannot be accurate and complete if these conditions are not fulfilled. Science is therefore most effective when the number of variable factors is small and when methods of measurement are highly accurate.

Consider the problem of investigating the way in which the length of a steel girder varies under natural conditions. Methods of measuring length are well-developed and present no problem. For practical purposes, the only variable which affects the length of the girder is its temperature. As the temperature rises the girder expands and becomes longer. As the temperature falls, the girder shortens. Accurate methods of measuring temperature are also easily available. It is therefore simple to investigate the relation between length and temperature. It is necessary only to take the steel girder, to measure its length and temperature, to heat it up to a number of different temperatures and to measure the length at each. The results may be plotted as points on a graph. The points represent experimental facts, namely the lengths of the girder at four different temperatures. However, it would be useful to know the length of the bar at temperatures in between those actually measured. If the points in this particular experiment seem to lie along a straight line it would be reasonable to draw such a line through them. Using this line, the length of the bar at any intermediate temperature can be determined without making any further measurements. This is known as intrapolation because, although it enables the length of the bar to be determined without measuring it, the process operates only within the range of temperatures and lengths actually measured. Intrapolation is usually a safe and reliable procedure provided that the experimentally determined points are not too far apart.

One might go much further. Since the points lie along such a straight line, why not extend the line beyond the range of actual measurements? It might then be possible to predict the length of the girder at a temperature far higher than any experimentally measured. This process is known as extrapolation because it goes outside the range of experimentally determined points. It may or may not be a reasonable thing to do. Unfortunately, in any individual case it is impossible to know whether or not extrapolation is valid. The line may continue straight as expected or may curve in any one of a number of different ways. In the case of the length of a girder extrapolation may be valid at one given temperature. But at a much higher temperature the bar might melt and the relationship between length and temperature become meaningless. This could not be predicted by extrapolation.

This is an extremely simple example with only one variable factor, temperature, to be considered. Both the phenomenon under study, length, and the variable can be precisely measured. In practice it is rare for all the variables which influence a phenomenon to be both known and susceptible to precise measurement. Even in elementary physics there is usually more than one variable. For instance, in an earlier example it became apparent that in studying the space occupied by a given amount of gas it is not enough to measure only the temperature. The pressure must also be noted. Before studying the phenomenon accurately one must be aware that both pressure and temperature can influence volume and one must have accurate instruments for measuring temperature and pressure.

Of course it is important to realise that scientific laws which deal with length, volume, temperature, pressure or any other subject are merely descriptions of what happens. They are not explanations. Boyle's Law, for instance, states that if the temperature of a given mass of gas is kept constant, the relationship between the space it occupies and the pressure can be described by a straight line. The greater is the pressure, the smaller is the volume. The law states that if the temperature is *x* and the pressure is *y*, then the volume must be *z*. It certainly does not explain *why* the volume should be *z*. It simply states that given those conditions this must be true. Like all scientific laws, it is an explanation only in the most superficial sense. It is important not because of the validity of its explanation but because of the usefulness of its predictions.

Although experiments in classical physics, chemistry and engineering can certainly present difficulties, these are not usually insuperable. These are the most reliable of all the sciences and by far the most successful. In biological science, in any science which concerns itself with life, the situation is much more chaotic. There are many more variable factors to be taken into consideration. We are not even aware of the existence of many of these, most of the known ones are poorly understood and accurate methods of measurement are available in only a few cases.

Suppose I am interested, as I happen to be, in the control of blood pressure in man. My first problem is one of measurement. The method of inflating a rubber cuff around the arm is quite satisfactory in routine medical practice. Unfortunately, for most research purposes it is quite unsuitable. It is not accurate enough and it cannot be used to give a continuous record of pressure under different conditions. The only way to measure arterial pressure accurately and continuously is to insert directly into an artery a needle connected to an electronic pressure recording device. This is an unpleasant and sometimes painful procedure which not even the scientists engaged in the research are prepared to go through too often. It is certainly unsuitable for routine use on patients and volunteer subjects. Even if a simple, safe, reliable and painless method of measurement could be devised, work on human beings would have to be strictly limited in scope. In the human I cannot cut a nerve to see what effect it has on blood pressure control, nor can I blithely inject some new drug. If one believes that the control of blood pressure must be investigated because of its importance in human disease, it is essential to begin an attack on the problem by using animals.

51

My willingness to use animals for research is an excellent example of survival logic. I start from the premise that everything should be done to ensure the survival of the human race and that it is legitimate to sacrifice animals in this cause. I have the utmost admiration for those few saintly individuals who believe that it is unworthy of man to sacrifice animals to save himself, provided that they realise fully the implications of what they believe and provided that they are prepared to renounce forms of medical treatment which have been developed as the result of animal experimentation. This, of course, means rejecting virtually the whole of modern medicine. I am a little contemptuous of those who make a very loud noise about animal experiments but who have no hesitation in rushing to the doctor as soon as their little toe begins to ache.

There are important limitations to the value of animal experiments. Right at the beginning a variable factor whose significance is unknown is introduced. I want to understand the control of blood pressure in man. But most experiments on man cannot be justified on ethical grounds and so I must use animals. I hope piously that the control mechanisms in man will be similar to those in cats, in rabbits, in dogs and in chimpanzees. But I can never be absolutely sure and I have brought in a variable whose importance I cannot easily measure.

Once I have decided which type of animal to use, I am faced with another choice. Shall I work with anaesthetised or un-anaesthetised animals? Most scientists, myself included, find the latter type of experiment excessively repugnant if there is a risk of causing pain. In Britain, at least, this natural repugnance is reinforced by strict legal control. The types of experiments which can be carried out on conscious animals are quite rightly very limited. Suppose, however, that I do want to investigate the effect of a particular drug on the blood pressure of the conscious rabbit. The central artery in the rabbit's ear can easily be seen through the skin. A transparent plastic capsule with a thin rubber diaphragm is slipped over the ear. The pressure within the capsule can be altered and its magnitude measured. The capsule is blown up until blood flow in the artery ceases. The pressure is then lowered slowly until a spurt of blood just overcomes it with every heart beat. In this way changes of blood pressure in response to the drug can be measured without causing pain to the animal.

Unfortunately, any animal's blood pressure is altered by its surroundings, by excitement and by a host of other factors. Most of these variables are not understood and their importance differs from individual to individual. There is no way of measuring them. Hence it is impossible to be sure that they are being kept constant while I deliberately alter the variable I want to study, the dose of the drug. Even if I make every effort to standardise the environmental conditions in which the measurements are made, it is likely that different animals will respond differently to the same environment. All that I can do is to repeat my experiment a large number of times, hoping that the response which occurs most frequently represents reality. In this way I hope that chance variations in the environment, such as traffic noise or unusual smells, will not have an undue influence. I also hope that by repeating the work a large number of times on many animals I shall avoid the misleading results which could arise because of some unusual individual, such as a rabbit which is excessively excitable. Despite these precautions, the measurement of a factor such as arterial pressure in a living animal can never yield very certain results.

The use of anaesthesia is one approach which aims to eliminate some of the variables which are in play when an animal is conscious. In fact, if I use anaesthetised rabbits I am little better off. I have merely replaced one set of variables by introducing another which I do not understand and which I cannot measure in any reliable way. In no single case is the mode of action of an anaesthetic understood. We are almost totally ignorant about the way in which these universally used drugs operate. One of the few things that is known is that anaesthetics themselves affect blood pressure. The size of the effect differs markedly from individual to individual and from anaesthetic to anaesthetic.

Similar uncertainties arise whenever any scientist attempts any experimental work on any living thing. It is therefore not surprising that the results of biological work are always thoroughly unreliable. As a result, in comparison with physics and chemistry, biological science seems slow and ponderous and tends to waste its time on trivia. Very many biological projects are carried out not because they are important but merely because they are possible. There is nothing wrong with choosing soluble problems. As has already been pointed out, one gets no prizes for failing to solve problems which it is not possible to tackle properly in the present stage of knowledge and of instrumentation. Yet it does not seem to me worth while to tackle something merely because it is possible if the subject is not also important. Far too few scientists in any field actually think about their research. Instead of struggling with significant issues, they find it much easier to pour out vast numbers of research papers by tackling trivial problems using well-established methods. Since professional advancement depends on the number of papers published rather than on their quality, concentration on insignificant but easily studied topics is obviously a wise practical course for any young scientist. But science might advance more rapidly, and the overwhelming tide of idiotic publications might recede a little, if research workers did more thinking and less experimenting.

Pure and Applied Science

In Britain, and to a lesser degree in other countries, there is a curious distinction between pure and applied science. The pure scientist carries out research for interest's sake alone, unsullied by any commercial motive and with no hope of a practical application for his work. The applied scientist devotes himself to practical ends. In Britain in particular pure and, by self-proclamation, useless science has acquired a peculiar prestige. As a result those with the best scientific brains tend to become pure scientists. Most persuade themselves that this is because they want to investigate the nature of the universe in an atmosphere untainted by commercial gain. They feel that only in this way will they have freedom to allow their minds to rove at will. In theory they may be right, but in practice I think that the factors which drive a bright undergraduate into pure science are much more simple and much more crude. First of all, his teachers are likely all to be pure scientists who know nothing about applied science and technology. The undergraduate can hardly help but remain ignorant of these grimy fields. He also sees that most of the people who go into industry are those with poor second-class or third-class degrees. Those who get firsts and good seconds continue as research students and most hope to become university lecturers. The bright undergraduate feels that he wants to be among his peers in an environment where he can test and prove himself against other high calibre brains. He does not want to go into an institution where most of his fellow workers will, by his standards, be intellectually dull. Inevitably he chooses pure science rather than technology.

Most scientists' lives are dominated by the problem of raising money to finance their research. A young man beginning his career needs the money to purchase the equipment to do the experiments which will build his reputation. An older man, a well-established scientist, may find that the size of the sum he can raise, the number of glittering knobs he can twiddle and the rank of the person he can persuade to come to open his new department are all useful props to the personality. Not surprisingly it is often difficult for pure scientists to raise money from government, from industry and from charitable institutions in order to finance work which is claimed to be without practical application. In consequence most pure scientists have developed a nice line of patter which demonstrates conclusively that most of the important advances in technology have stemmed from pure men tinkering without commercial motive. There are many well-known stories. One is that of Faraday, who in 1821 developed the principle of electromagnetic induction. On this principle depends the working of all electric motors and dynamos. All modern industry is based upon it. Faraday actually constructed a crude electric motor and was demonstrating it when a lady asked, "Of what use is that?" Faraday is reputed to have replied, "Of what use is a new born baby?"

Another favourite is the story of Hertz, who in 1886 discovered the type of radiation which we now know as radio waves. Hertz himself described his discovery as utterly useless.

Science is God

It was left to Marconi to see the potentialities. Even when radio had been demonstrated, the pure scientists said that it would be useless over long distances because the waves travelled in straight lines. Because of the curvature of the earth the waves would shoot off into the atmosphere and be undetectable at any significant distance from the transmitting point. There was no doubt that these men had their facts right. The argument seemed conclusive, but Marconi ignored it for no very good logical reason. He just believed that the pure scientists were wrong. Like an idiot he tried to send radio waves across the Atlantic and he succeeded. Marconi's dream had been saved by a factor of whose existence both he and the pure scientists were unaware. There is in the upper regions of the atmosphere a layer which reflects radio waves. Marconi's waves were reflected back and forth between this layer and the surface of the earth. In this way the curvature of the earth was overcome. This is the most salutary of warnings about the need for caution in accepting the apparently watertight arguments of learned men.

The blarney of the pure scientists certainly sounds convincing to the uninitiated. Many commercial firms have fallen for it and as a result are supporting pure research both in their own laboratories and by means of grants. In my view they are quite mistaken to provide this support unless they are doing it for philanthropic rather than for commercial reasons. Pure research as a commercial proposition is a waste of time and money except in so far as it may attract high calibre brains into the firm. There are the odd startling stories but they are only the exceptions which prove the rule. In the main pure research is what it claims to be, pure and useless. Even when it does turn out to be useful, the applications are likely to be found in some totally unexpected field which is quite unrelated to the commercial aims of the people providing the financial support. Furthermore, the commercial usefulness of pure research is rarely apparent at the outset as the stories of Faraday and Hertz show. As a result, the fundamental work will almost certainly be freely published in scientific journals long before its applications become apparent. The vital information is therefore likely to be just as available to a company's main competitors as to the organisation which backed the project. Faraday and Hertz certainly did not collect the profit from their discoveries.

In spite of all this, it would be a pity if industry stopped supporting pure research, because in the end technological advances do depend on an understanding of the fundamental properties of the material of which the universe is made. However, I believe that it is deceitful to suggest, as many scientists do, that in the long run pure research is a commercial proposition which will more than pay for itself with increased profits. Except in the very rarest cases, pure research will not help a company's Stock Exchange rating one little bit. It is more likely to be an expensive albatross. Those industrialists who do finance pure research must realise that they are making a philanthropic gesture and are not investing in something which is likely to bring in any commercial return.

In general, the quickest and best way to get an answer to a particular problem is to pay people to think about that problem directly. It must never be forgotten, however, that science is like politics in being the art of the possible and of the soluble. It is worth paying someone who is very clever a great deal of money to give a down to earth opinion as to whether in principle a problem is likely to be easily soluble. If methods for tackling it are not immediately available, it is probably cheaper to abandon the scheme rather than to set out into an unknown field and try to develop new methods. It may well be much more sensible to wait until someone else has done the expensive hard work.

There is no difference between the intellectual processes involved in tackling pure and applied problems. Often the possibility of a practical application adds greater piquancy to the work of even a pure scientist. Many academics who would defend to their last breath the importance of useless research are childishly delighted when someone turns up with a practical application for their own pet baby. Both pure and applied problems may be tackled in two quite different ways, by routine chipping away using established methods or by inspired dreaming which in a moment turns a whole field of knowledge upside down and provides a brilliant and totally unexpected answer. Corresponding to these two approaches, the scientists themselves come in two fundamentally different varieties. These are not pure and applied as some pedants would like one to believe. They are men capable of only routine thinking and men capable of truly creative flashes.

Those who do the routine thinking do not need to be very intelligent. Their education is an exercise in mass production and they can be turned out in enormous numbers. They are the pure scientists who, having seen others isolate a particular protein in the rat, go on to isolate the same protein in the rabbit, cat, guinea pig, goldfish, human being and Mongolian frizzle-haired purple-striped baboon. They do not have an original thought in their heads but use an established technique to tackle a long string of problems which differ from one another in name only. They build up a large and impressive life list of publications. They may become important in the scientific world and even more so in government, but in reality they contribute virtually nothing to scientific progress. They do not even have the satisfaction of knowing that their work is useful. They have their counterparts in applied scientists who fiddle to make lipsticks more kissable or cough lozenges more palatable. At least the latter have the satisfaction of knowing that their work has a practical application.

In contrast, scientists capable of leaps of creative insight are as rare as great artists. No university, no industry, no government should fail to make every effort to find these men and to keep them working under satisfying conditions. These are the men who will provide radical new solutions to old problems. They are the Flemings who will use a laboratory accident to demonstrate the anti-bacterial properties of penicillin. They are the Floreys who will take that useless laboratory demonstration and turn penicillin into a life-saving drug. They are the Barnes Wallises who will develop a swing-wing aircraft and find that the worth of the idea is recognised only twenty years later and then by another competing nation. Any organisation which finds that it has one of these men must be prepared to fight to keep him and to support his creativity. Their efforts will be more than amply rewarded.

Mathematics and Science

There is much misunderstanding about the role of mathematics in science. Most lay people and many scientists seem to believe that science is largely mathematics. This is not true. Mathematics bears the same relation to science as do the rules of logical reasoning. In effect, that is what mathematics is, a system, or rather a multitude of systems of logic. Given certain premises, mathematics allows certain conclusions to be drawn. Mathematics can guarantee only the logic of the progression from premise to conclusion. If the premises are faulty or unreliable, no matter how elegant or brilliant the mathematics, the conclusions must also be faulty and unreliable.

A simple illustration may make this point more clear. I go into a store to purchase a set of wine glasses. I want ten of them. After careful thought I choose a pattern and the assistant tells me that each glass costs ten shillings. An elementary piece of arithmetic then tells me that the total cost of my purchase will be 100 shillings. Given that I want ten glasses, given that each glass costs ten shillings, my mathematics demonstrates with perfect accuracy that I shall have to pay one hundred shillings in all. Unfortunately, the assistant made a mistake. The glasses should cost eight shillings each and not ten. Despite the reliability of my mathematical reasoning, my estimate of the cost is wrong because the information on which it is based is faulty.

In rather more complex forms, similar situations occur again and again in science. A complex experiment is performed and the results are not very reliable. It is impossible to ensure their repeatability. The mathematical scientist gets to work on this raw experimental information. He carries out a series of brilliant mathematical manipulations and produces his conclusion. Both the scientist himself and his readers are dazzled by the elegance of his work and the conclusion becomes one of the established facts of science. Everyone forgets that because the original data are unreliable, so the conclusions must be unreliable, irrespective of the brilliance of the mathematics. Only when some iconoclast bothers to look back at the original description of the experiment is the mistake revealed.

At the moment, like several other countries, Britain is much concerned about the so-called "swing from science". Science and technology are decreasing in popularity amongst school leavers. The proportion of scientists to non-scientists in universities is becoming smaller and smaller. The British Government, very worried by this trend, appointed the high powered Dainton Committee to investigate it and to suggest how it might be counteracted. They recommended that an attempt should be made to make everyone at school study mathematics until they left. A brilliant and fatuous idea! I personally can think of nothing more likely to accentuate the swing. The ability to cope easily and well with mathematics is as individual as the ability to play the flute or the ability to become a brilliant tennis player. Most people, even in the sixth form, are frightened of complex figures, and the new maths we hear so much about is not likely to make very much difference at that level. Any subject can be taught attractively only if the teacher fully understands it. This is perhaps more true of mathematics than of anything else. Yet, as everyone knows, there is a devastating shortage of good mathematicians in school. This is not surprising: they can get much better pay and working conditions elsewhere. If mathematics is to be taught to everyone in the sixth form it will inevitably for the most part be taught drearily by third-rate teachers who do not understand it. Any lively and intelligent sixteen year old will turn with relief to the delights of history or English literature taught by a graduate in the subject. If one wishes to turn the swing from science into a landslide, by all means insist on mathematics being taught to all sixth formers.

There is a more fundamental objection to the Dainton Committee's report than the crude practical one of teacher supply. It is undoubtedly true that a knowledge of advanced mathematics is useful to any scientist. So is a knowledge of German, of Russian, and of Japanese. So is a manual skill in the manipulation of lathes and drills and experimental hardware. Yet none of these things is absolutely essential. In many scientific fields, and in particular in biology and medicine, one can get along perfectly well with nothing more than an Ordinary level G.C.E. knowledge of mathematics, something which almost all sixth formers already possess. Without doubt it is still possible to be a great biological scientist without being a mathematician. It will remain possible to be so for many years to come provided, of course, that high-powered committees do not impose conditions which say that a man must have a good knowledge of mathematics before he can be allowed near a biological research laboratory. Biology is by no means so advanced and so dependent on mathematics as physics was in the first part of the nineteenth century. Yet at that time, Michael Faraday, whose mathematical ability was utterly negligible, became one the great physicists of all time because of his vivid imagination and his skill in experimental fiddling. Of course, today Faraday would never have come near an undergraduate course in physics, let alone a physical research laboratory. But he would have been excluded not because he lacked skill in physics but because he lacked mathematical ability. I hope that we never reach the stage where we exclude potentially great biologists from undergraduate courses in biology merely because they have not studied mathematics at advanced level. What biology needs today is people with imagination and not ones with mathematical skill. Major advances in biology and medicine will continue to be made by people whose mathematics does not extend beyond the simple arithmetic level. Therefore, if science is to attract more bright young sixth formers, it is essential to make it clear that in many scientific fields no advanced mathematics is required. The areas where this is true are largely those which deal with life. It is here that our understanding must be increased if we are to come to terms with living on our overcrowded planet without destroying everything we hold most dear. What is needed in a biologist or medical scientist is not an intimate knowledge of the intricacies of calculus but a lively imagination, an understanding of the fundamental principles and uncertainties upon which science is based and a willingness to submit creative dreaming to judgment by an objective higher authority.

The Motivation of Scientists

Scientists are driven by much the same motives as any other group of professional men. Some few, a very few, do it only for the love of the chase, for the joy of grappling intellectually with nature, for the thrill of answering some of the problems posed by matter and by life. Such men are indifferent to recognition or to financial reward and ask only for conditions in which they can carry out their work in peace. A larger, but still small group, adds international recognition to the love of the chase. They want to make important discoveries and they want other workers to recognise the power of the brain that made them. A third group adds financial reward to intellectual excitement and international recognition. With all three, it is the science that matters: other things are of secondary importance. People driven by this demon need no sticks and carrots to persuade them to work. The difficulty for their families and friends is to stop them driving themselves into the ground. I know several such men and they may perhaps best be summed up by a Christmas card sent to one of them by his wife and children. It read, simply, "From the people with whom you eat Sunday lunch". These individuals are the salt of the scientific earth, but girls should be wary of marrying them and children should not expect too much of them as fathers. Nevertheless, because of their unlimited appetite for work and because of their success in research they should be sought by every university, every company and every government department. The man who lives only for science may be a pain in the neck to those who work with him and a severe trial to his family and friends, but to the organisation which employs him he is a treasure not to be set aside lightly.

Most scientists, it need hardly be said, are not like this. They are interested in science and they enjoy their work, but their job is a job. Joy in discovery and even international recognition matter little besides their income and their status in the immediate society in which they live. Like most men, they live primarily for their families and for what the money they earn can buy in their leisure time. Scientists of this type often fall into two further classes. The one is not committed to science as a life career. He is happy to move away into management, where he will be paid more and will have more status and influence in the organisation. The second type, even though science is not the most important thing in his life, wants to earn his living only as a scientist. He is lost outside the laboratory and loathes the thought of becoming an administrator. Yet at the age of 35 he sees his fellow management-inclined scientists moving rapidly ahead of him in income and status. What is even worse, he sees arts graduates who may be far inferior to him in intellect going in the same direction. Ten years previously, at the age of 25, he had just got his doctorate. A stream of callers from industry beat their way to the door of the university laboratory where he worked, plied him with lavish dinners, pleaded with him to join their firms and offered him all sorts of fringe advantages. He could afford to pick and choose among a dozen suitors. For a man of his age he became very well paid. Now, ten years later, he sees with uncomfortable clarity that if he is not to become stuck in an incomes backwater he cannot afford to stay in science. For the sake of his family and of his own self esteem he must make the effort to wrench himself from his beloved laboratory. He does so, gains an increase in salary, does ineffectively what he regards as an unsatisfying job and spends the rest of his working life being miserable.

This large group of scientists, forced into administration by economic pressures, represents a colossal and frivolous waste of talent. In Britain, as undergraduate and postgraduate students, these men have been educated for six or seven years at the nation's expense. They work for five or ten years as scientists for their company or in the Civil Service. They are just becoming really useful when they realise that unless they are prepared to leave science behind they will never attain the income level which their arts contemporaries may expect. They then may spend thirty years in misery, making the best of a bad job for which they have not been trained. In an age when almost everyone seems to be screaming about the shortage of scientists, is it not stupid to train a man for science and then to use him for a period only about equal to the length of his undergraduate and postgraduate course? Industrialists and governments who moan about the lack of scientific manpower are wasting thousands of trained men in unproductive posts. This is because they short-sightedly believe that a desk-bound administrator must be paid more than a man who wants only to dirty his hands in practical laboratory work. A man who wants to do science for the whole of his life should be encouraged to do so and should be paid for it as handsomely as his management colleagues. Overnight, without increasing the size of a single educational institution, the effective output of trained manpower would be increased two-, three- or even four-fold. The problem of scientific education is not only that too few scientists are being trained but also that too many are being lost to more lucrative lines of employment. If we simply paid older scientists adequately many problems would be solved without the need for a tremendous expansion of scientific education.

The main argument against this policy is the supposed loss of scientific ability which is believed to occur after the age of about 35. In fact, there is very little evidence that this is true except possibly in mathematics and theoretical physics. In the biological sciences, in medicine and in engineering, the feel of the subject and intuition as to just what can and what cannot be done matter a great deal. Experience is invaluable and a scientist in these fields does not fully mature until he is 35 or 40. Relatively few great medical or biological discoveries have been made by younger men. Indeed Pasteur, perhaps the greatest biologist of them all, did much of his best work after the age of 50 and after he had suffered a major stroke.

One possible explanation for the myth of the superiority of the very young scientist is that the young man is more hopeful and more enthusiastic. He is not inhibited by the difficulties of which the older man, who has been in the field for many years, is all too conscious. But this boredom, this ineffectiveness and this sense of being overwhelmed by the problems are probably dependent more on the length of time for which a man has been working in a narrow field than on the age of the man himself. Set such a middle-aged or elderly man to work on a quite different problem in another field and he may again become hopeful and effective. I personally find that when I direct my attention to a new subject, all my good ideas are formulated within about three months of starting. It may take a long time for the ideas to be followed up, I may modify them in minor ways, but rarely do I hit upon anything radically new after this time. Although as yet I have had relatively little experience, I should guess that about five years may be the optimum length of time to spend in any one narrow field. Only the very beginning of this period is likely to be truly creative. The remainder will be spent in devising experiments, in persuading them to work, in proving one's hypotheses or in demonstrating them to be invalid. After this order of time lapses, more and more energy tends to be directed at trivia and a feeling of boredom and despair creeps in. The secret of scientific youth is to keep changing one's problem. Creativity does not fall off much with age, but enthusiasm for any particular project certainly does.

Perhaps the arch example of my point of view is that brilliant applied scientist Barnes Wallis. This is a man who has continually sought new problems and new outlets for his intellect. After earlier distinguished work on airships and conventional aircraft, in his fifties he produced the bouncing bomb, in his sixties he designed a swing-wing aircraft and in his eighties he is still pouring out a stream of stimulating ideas. Sir John Eccles who recently won the Nobel prize for medicine is another example. He devoted what many would call the best years of his scientific life to demonstrating that a particular hypothesis of nerve cell function was true. In his forties he was forced to admit that his opponents were correct. Most men in his position would have retired to some quiet chair of physiology and vegetated. Eccles continued to do active research, this time devoting all his efforts to investigating the implications of his opponents' work. His output during the past twenty years has been prodigious and shows little sign of abating. In 1963 he was awarded his Nobel prize. So much for scientists being finished at 40!

I cannot resist ending this chapter with a brief Eccles story which has really no relevance. Sir John is known for his forthright views, for his dominating personality and for his fecundity as a Roman Catholic father. Some years ago he was given the great honour of being made a Vatican Knight for his services to science. On hearing this one competent but much less flamboyant and less successful worker in his field was heard to mutter, "Good heavens, does that make him infallible?"

Chapter 5

The Uncertainty Principle

In the last two chapters, various practical considerations which limit the reliability of scientific conclusions were outlined. In biological science in particular, so many variables influence any measurement that it is impossible to control them all and even impossible to know of the existence of many of them. However, this is a practical objection and one which in theory could be surmounted. This chapter is concerned with a quite different form of limitation. It is one which even theoretically cannot be avoided and which strictly sets the bounds within which the validity of scientific work can be assured. This limitation arises from the fact that objective measurement lies at the core of science. Unfortunately measurement can be objective only under rather stringent conditions.

Before any measurement can be made, it is essential to have a standard unit for comparison. Length is measured in terms of feet or metres, weight in terms of kilogrammes or pounds, volume in terms of pints or litres. The object under study must be observed and compared with such standards. No one argues about how long a centimetre should be or how heavy is a pound. If I say that a steel bar is ten centimetres long people anywhere in the world will understand what I mean. In classical physics and engineering such units have been standardised for many years and have gained universal acceptance. There is no problem of communication between scientists working in different laboratories and in different parts of the world.

A further essential feature of any method of measurement is that the act of observation should not in itself alter the state of the object being observed. Again this condition is usually easily fulfilled in classical physics and engineering. When measuring the length of a steel bar, for example, light must be used in order to compare the bar with the ruler or the tape measure. The bar is observed by illuminating it and by using the light reflected from it to the eye. The fact of shining the light on the bar in no way alters its length. We can be sure of the reliability of the measurement because the method of observation does not alter the length of the bar.

If measurement is to be accurate and if we are to be able to rely absolutely on it, other conditions must also be fulfilled. For example, it is essential to be quite sure what one is measuring. That may sound silly but it is a very real problem. It may be illustrated by one commonly used method of chemical analysis. Many measurements of chemical concentration depend on the fact that the substance under study reacts with another test chemical to give a particular colour. Suppose that I want to measure the concentration in blood of substance X. I know that when X reacts with a test chemical, Y, a blue colour develops whose precise intensity can be measured by means of an instrument known as a colorimeter. The higher the concentration of X, the more intense is the colour which develops. By measuring the intensity of the colour it is therefore possible to estimate the concentration of X in the blood.

It is apparent that if this method of measurement is to be accurate, one important condition must be fulfilled. There must be no other substance in the blood apart from X which reacts with Y to give a colour. If some other substance is present the result in meaningless. Unfortunately, especially when working with biological material such as blood or urine, it is often impossible to be sure that this condition is fulfilled. The literature of biological science is littered with the false observations of men who have measured not only what they thought they were measuring, but a number of other extraneous substances as well. It is not stupid to say that one must be quite sure what is being measured.

Another requirement, if objective validity is to be ensured, is that the measurement must be repeatable. Only if an observation is repeated a number of times and only if the same result is obtained each time is it possible to be reasonably sure of reliability. Single observations of unique events are scientifically valueless. It is impossible to be sure that a single measurement is unaffected by some technical error or chance variable. The making of only a single observation would be rather like testing shooting ability on the basis of only one shot. Given one hundred shots, a brilliant marksman might hit the bull ninety-seven times, while a poor one might score only thirteen hits. There could be no doubt as to who was the better marksman. But if only one shot were allowed, the good man might miss and the poor marksman by chance might score one of his rare successes. A quite erroneous assessment of relative ability would then be made.

Series of observations by a single man are more reliable than an isolated observation by that man. But series are still subject to bias, especially if the scientist has any preconceived ideas as to what the result should be. It must be understood that bias is quite distinct from dishonesty. Dishonesty is a deliberate attempt to falsify the results. Bias is a quite unconscious falsification which creeps in because the observer thinks he knows what the result should be. Bias merely means that its perpetrator is a human being and does not indicate that he is also a criminal. It can work both ways. The man with the pet hypothesis may be so anxious to prove its validity that unconsciously his measurements tend to give the results he requires. On the other hand, he may be so anxious to ensure that the results are honest that unconsciously he pushes the observations in a direction which goes against his own hypothesis. The only way to get round this is either to avoid preconceived ideas or to have several series of observations made by men who have quite different views as to what the results ought to be. Bias has proved to be particularly troublesome in medicine when assessing the value of new drugs. It is often impossible to reduce any change in a patients condition to cold figures, and the doctor must rely on his own impressions of such things as the intensity of a pain. When testing a new drug it is essential to compare the treated patients with a control group of similar patients who receive dummy pills (placebo). Obviously the patient himself must not be aware of the fact that he may be receiving dummies. It has become apparent that the doctor, too, must not know which patient is receiving the real drug and which the placebo. If he does know, according to what he feels about the treatment, he makes a more or less favourable report about the treated group than reality would warrant. Only if both the patient and doctor do not know who is receiving the real pills and who the placebos can the doctor assess progress reliably. The decision as to who is to receive what must be made at random by someone such as a secretary who has no direct connection with the medical part of the trial. She must keep all the records, and only after the trial is over must she reveal who was receiving what. This is known as the principle of the double blind trial, and doctors have learned to be extremely wary of using drugs not tested in this way.

Finally, anyone who wants to be certain of the validity of his work must take pains to ensure that he is not playing what has been called the substitution game. Suppose one wants to measure something but that no reliable method is available. Blood progesterone level (see chapter 3) is a good example. It is relatively simple to measure the concentration of the destruction product of progesterone, pregnanediol, in the urine. This one does, believing that the amount of pregnanediol put out in the urine will reflect the amount of progesterone in the blood. But when reporting the results, one does not only say that in toxaemia pregnanediol levels in the urine are low. One says also that blood progesterone levels are low even though these levels have not actually been measured. It is this second observation that passed into the textbooks and into scientific folklore. The real measurement is of urinary pregnanediol. But this is not particularly interesting and so in discussing the results, blood progesterone is quietly substituted for it. In the original work the connection between the two may be apparent, but gradually it is obscured and eventually everyone forgets that the well-known fact that progesterone levels are low in toxaemia is not based on progesterone measurement at all.

Another good example is that of the susceptibility of patients to the development of coronary heart disease. The cause of coronary disease is as yet unknown, although certain patterns of behaviour such as smoking, over-eating and lack of exercise are undoubtedly associated with it. There is some hotly disputed evidence that those with high cholesterol levels in the blood are much more likely to develop the disease. Now the cholesterol level is easy to measure accurately but susceptibility to coronary disease is not. What more natural, therefore, than to equate the measurement of cholesterol levels with the measurement of susceptibility to coronary thrombosis. What more natural to say that diets and drugs which lower the blood cholesterol level also reduce the susceptibility to the disease. It may be natural, it may even be true, but in the present state of knowledge it is an unjustified substitution. As will be seen later, nowhere is the substitution game played more effectively and more dangerously than in the study of the human mind and in those fields of research optimistically and misleadingly called the social sciences.

In summary, if measurement is to be reliable and objective, the scientist must use standard units, he must know what he is measuring, he must design his experiments so that they are repeatable and so that bias is eliminated, and he must not play the substitution game. Perhaps most important of all, he must be sure that the method he uses does not in itself alter the thing which he wishes to measure and that it has an objective validity which does not depend on who makes the observations. In some fields, and in particular in engineering, chemistry and classical physics, these problems are relatively easily solved. It is significant that these sciences are also the ones which have been most successful.

Heisenberg's Uncertainty Principle

Suppose that I wish to plot the flight of a bullet or of a billiard ball. One way in which I might do this would be to take photographs using a high-speed camera and powerful lights. I can be confident that the bright light will in no way alter the flight of the bullet or the direction in which the billiard ball rolls. The particles of light are so tiny in comparison to the relatively huge masses of bullet and ball that they cannot significantly impede progress. I can be reasonably sure that the fact of making the measurement in no way alters what I want to measure. The same results would be obtained by any technically trained observer, black or white, male or female, communist or capitalist.

Problems can and do arise even in physics, chemistry and

engineering. A particularly simple and easily understood example is that of temperature measurement. If I want to measure the temperature of my child's bath, an ordinary mercury-filled household thermometer will do very well. When the thermometer is dipped into the bath, heat flows from the water to the mercury and the temperature of the latter rises until it is equal to that of the water. As the mercury is heated it expands and the column of silvery liquid rises up the thermometer tube. The amount of heat required to warm up the tiny blob of mercury is far too small to alter significantly the temperature of the water. The fact of making the measurement does not alter to any noticeable degree the temperature of the bath water. In contrast, suppose that I wanted to measure the temperature of a tiny drop of water at the bottom of a test tube. The household thermometer would be quite useless. When the bulb came into contact with the water droplet, the heat flow from one to the other would seriously alter the temperature of the water. The measurement would be meaningless. It could be made meaningful only if a thermometer were dévised which required so little heat that the temperature of even a droplet of water would not be significantly altered. This is not too difficult to do. In principle, all problems of measurement in classical physics and engineering can be easily tackled.

In the nineteenth and early twentieth centuries, science dealt almost exclusively with problems in which the difficulties of measurement were relatively few and strictly practical. Measurements could safely be made without the observation itself altering the phenomenon under study. The success of the scientific method led to euphoria and over-optimism among many educated laymen and also among scientists who ought to have known better. Science was the hope of the future, the ultimate answer to all the problems of mankind. There was little reason to suspect that there were any realms of study in which the problem of measurement was anything more than practical. There seemed no reason why science should not investigate with absolute certainty everything from the infinitely large to the infinitely small. All seemed merely a matter of time and technical juggling.

With the development of nuclear physics it became apparent that the certainty might not be so certain after all. The ancients believed that the universe was made up of four elemental substances, earth, air, fire and water. All material consisted of mixtures of these elements in different proportions. In fact we now know that there are about a hundred elements and that earth, air, fire and water are not among them. Some of these elements are common and have familiar names: gold, silver, carbon, oxygen and lead. Others, like niobium, technetium, rubidium and caesium are rarer and more exotic. Suppose I take a lump of any element, say gold, and cut it into smaller and smaller pieces. If I had an instrument fine enough, eventually I would reach a particle of gold which could not be cut further and remain gold. This particle is an atom of gold. If split, it would break up into numerous sub-atomic particles, the protons, neutrons, electrons and other varieties of which the atoms of every element are made. The nature of each atom depends on the number of each type of sub-atomic particles which it contains. By manipulating the numbers of these sub-atomic particles, the nuclear physicist has realised the old alchemical dream of the transmutation of elements.

Nuclear physics is the science which deals with these almost inconceivably minute particles of matter. And that is the problem. How is it possible to make measurements of the behaviour of such objects? In order to photograph a rolling billiard ball we direct at the ball a beam of light consisting of the minute particles known as photons. Because they are so small relative to the size of the ball, the photons do not influence the ball's course. They are reflected from it and the reflected photons are detected by the film in the camera. The ball can be observed and its position and speed accurately measured because of the way in which it reflects the photon beam. Sub-atomic particles are observed in a similar way. Beams of other minute particles are directed at them and are deflected on impact with the particle under observation. The presence of the particle under observation is detected as a consequence of the way in which it deflects the course of the particles used to make the observation. But in nuclear physics the problem is that the particles used to make an observation are comparable in size to the particles being observed. It is possible to plot accurately the course of a billiard ball only because the photons are so incredibly tiny compared to the size of the ball: they could not possibly alter its course. But in nuclear physics the beam of particles used to make the observation inevitably itself alters the position and velocity of the particle being observed. The difficulties were summed up by Heisenberg in his famous Uncertainty Principle formulated in 1927. This effectively says that it is impossible even in theory to know everything about the physical state of one of these minute particles. If we know the speed with which it is travelling we cannot at the same time know its position. If we know its position we cannot know its velocity. The making of the measurement alters the state of the object being measured. There is a barrier to knowledge beyond which it seems impossible to go, not for any practical reason but for a theoretical one which cannot be avoided. Man cannot know all there is to know about the atom and its constituents.

The Biological Uncertainty Principle

Heisenberg's Uncertainty Principle is well known, is much discussed by philosophers and has a negligible effect on the practical operation of our lives. At the other end of the scientific scale is another uncertainty principle which also cannot even theoretically be surmounted. It may have enormous practical consequences, yet it is poorly understood and studiously ignored by those it most affects. Heisenberg demonstrated that man cannot observe the fundamental particles of matter without so changing them in the process that he can never know what was their physical state before he made his observation. At the other extreme, man can never reliably observe man without the fact of making the measurement altering the state and behaviour of the human being under study. This biological uncertainty principle means that much of the activity which calls itself science is really nothing of the kind. The proper study of mankind may be man, but most of such study lies outside the bounds of science and must never be misleadingly given that name. The so-called social sciences, for example, are about as far removed from science as is witchcraft from high precision engineering.

The problem begins with the study of man's body. Every medical student knows some variant of the classic story of the interaction between observer and observed in the course of research on blood pressure. A young man volunteered to come along to a laboratory each week to act as the subject for experiments on the control of arterial pressure. It soon became apparent that the sessions were falling into a pattern. One week the level of pressure would be steady and relatively low. The next week the pressure would be significantly higher with a highly irregular set of recordings. The pattern repeated itself for several weeks and the doctor in charge of the experiments could make no sense of it. He wondered whether he might have discovered some hitherto unknown male cycle corresponding to the menstrual cycle in females. Then one night in a restaurant the answer came as he saw the young man eating out with Elizabeth, one of the girl laboratory assistants. The doctor checked his records. Every week when the pressure had been high and irregular Elizabeth had been assisting. Every week when the pressure had been low and the recording stable another girl, Janet, had been helping. This is a trivial example but the principle applies to any work which involves human beings. One cannot observe a man without altering him in the process.

In medicine, it is usual to assume that the doctor himself has little effect on the outcome of a course of drug treatment. When two methods of treatment are compared, it is the drugs and techniques and not the doctors which are assumed to be different. Recently, however, some have begun to wonder whether the personality of the physician might have a great deal more effect in the treatment of physcial disease than had hitherto been supposed. Some of this work may be found reported in *Controversies in Internal Medicine*, edited by Ingelfinger.[5] In order to make the principle intelligible, the following account has been much simplified, but it is based upon a real experiment. Four doctors were working on the problem of rheumatoid arthritis.

They all used and believed in precisely the same form of treatment. They decided to set up a trial in order to find out whether the personality of the doctor had any effect. Dr. Brown selected the patients for the trial. He endeavoured to ensure that the disease was of similar severity in all of them. Miss Jackson, a secretary at the hospital totally unconnected with the medical problem, allocated the patients at random to the three other doctors, Smith, Jones and Wilson. Dr. Brown had no idea which patients went to which one of the other three. Drs. Smith, Jones, and Wilson then treated the patients and saw them regularly over a period of about six months. At the end of this time, Dr. Brown again examined them and assessed their progress, still without knowing which other doctor each had been attending. After all the information had been collected, the names of patients and doctors were linked. Although all three used exactly the same methods of treatment, Dr. Wilson produced distinctly better results than the other two. This suggests that the influence of the doctor's personality is of considerable importance in the physical progress made by each patient.

If the influence of doctor upon patient during physical treatment is so great, it is certain that in any so-called scientific study of the physical attributes of human beings, the observer will alter the state of the person being observed. Unless the same problem can be studied independently by several different observers, and unless a carefully designed trial is carried out, it is impossible to assess the significance of the impact of the experimenter. Such trials, similar to the one which studied the effect of the doctor in rheumatoid arthritis, are laborious to organise and are quite impossible to carry out in the great majority of cases. There is neither the time, nor the technical expertise, nor the money available for such complex undertakings. Experiments therefore tend to be performed by only one observer with no cross checks to assess the impact of his presence. Rather than carry out a major piece of reliable research, many scientists studying man find it easier to churn out simple unreliable work and to hope for the best. With any luck they will be appointed to chairs before the inadequacies of what they have done are revealed.

Thus, in studying the physical attributes of man, we encounter an uncertainty principle similar to that enunciated by Heisenberg. The fact of making a measurement alters what we want to measure. It is impossible to know what the state of man would have been had the measurement not taken place. This biological uncertainty principle is an insuperable barrier to the full understanding of man by man. The effect of making the observation on the phenomenon observed differs from man to man and from observer to observer. The problem is that of man's consciousness. As soon as he knows that he is being observed, his behaviour changes. This would be true even if one withdrew the experimenter altogether and observed the man by means of machinery. There is only one way to avoid the problem. That is to devise a situation in which a man has no idea that he is being observed. Fortunately in practice it is virtually impossible to make repeatable, scientifically useful observations in such a situation. Even if it were possible, I believe that it would be unethical to set up such an experiment. No human being should suffer the indignity of being studied without being aware that he is being studied. No experiment is as important as that. It must be regretfully admitted that some who call themselves scientists have endeavoured to use situations in which a human being, usually a mentally defective one, has no idea that he is being observed. They have met with little success, but that is not the point. I believe that even the effort is morally wrong. Scientists must not be Lucifers who try to make themselves gods. The attempt to observe a man without his knowledge is an attempt to do just that.

All observations by man on man are therefore inherently untrustworthy. They lie outside the strict realm in which science is a valid and reliable method of investigation. If this is true of physical attributes, how much more true is it of the much more interesting properties of the mind. The problems of learning, of intelligence, of emotion and of motivation are the ones which the layman wishes to understand. He is not particularly interested in the control of arterial pressure, but he is avid for information on what governs the emotional life of human beings. As a result, in recent years there has been an explosion of activity in those subjects which deal with the behaviour of the higher parts of the nervous system, in psychology, psychiatry and psycho-analysis. These subjects are often dubbed scientific and because they are scientific they are thought by the layman to be completely reliable. Interesting they may be, useful too within defined limits, but they are not scientific and they are not reliable in the sense that classical physics and chemistry are scientific and reliable. I believe that the major reason for their lack of scientific validity is not the practical one that they are young subjects and that as yet little money and brain power have been expended upon them. I believe that they are comparatively unreliable because of the theoretically insurmountable problem of the impact of the observer upon the person studied. In human psychology, psychiatry and psychoanalysis, the attitude of the observer and the motives which lie behind the questions he asks are just as important in determining the results as is the mind of the human being observed. Both the experimenter and the experimental subject are unique individuals. The interaction between the two is a unique, unrepeatable event. There is no way of assessing the effect of the observation upon the observed person. This is bad enough. Even worse is that investigations of man's mind do not fulfil the other conditions for reliable measurement discussed earlier in this chapter. They do not use internationally accepted units, they are uncertain of what it is that they are measuring and they play the substitution game quite recklessly. Professor Thomson devises a test of intellectual ability. When Mr. Smith scores higher marks than Mr. Jones in this test, Professor Thomson does not say what he is entitled to say, namely that Mr. Smith performs better than Mr. Jones under the conditions of the test. The Professor says that Mr. Smith is more intelligent than Mr. Jones even though he has no idea of what intelligence is or of what his test is measuring. When an industrial psychologist carries out his tests on bright young recruits to a firm, he does not say: "My tests are inadequate and uncertain in operation but were carried out most effectively by Mr. Jackson." He says, "Mr. Jackson is excellent managerial material". When a Civil Service psychologist assesses an entrant for stability of character and suitability for secret work, he does not say "Mr. Phillips performed my test well", but "Mr. Phillips is stable and reliable". All this is not to say that the investigation of man's mind is always useless and utterly unreliable. It is merely to emphasise that such investigations, whilst valuable within their own limits, are not scientific. They do not, therefore, have the validity which the use of that word implies.

A particular example of the observation of man by man is the testing of intellectual ability. One might expect that after all these years such testing would have been made scientific and would have an objective reliability and validity. There are two fundamental reasons why this has not yet been done and why it will never be done. First of all, success in tackling an intellectual problem depends as much on the motivation of the tested subject as on that subject's mental ability. The person who is moderately keen and interested is more likely to be successful than either the one who could not care less or the one who is so desperate to succeed that the balance of both mind and body is disturbed. Secondly, there is no generally agreed definition of intellectual ability, nor are there any generally agreed methods of testing it. If there were generally agreed definitions and methods, there could be no guarantees that these would be valid and useful. Any examination is therefore based on the subjective decisions of the examiners as to what should or should not be tested. Any examination can test only the ability of a student to do that examination under those conditions and nothing more. Examinations must be of doubtful validity as assessments of intellectual ability and as predictions of the performance of an individual in later life.

The unreliability of academic examination is the strongest weapon in the armoury of those who campaign for the abolition of these mental gymnastics. They say, quite rightly, that an examination paper performed under conditions of stress is not necessarily a valid guide to performance outside the examination hall. However, as a believer in examinations, despite all their faults, I feel there are points which should be borne in mind. The first is that life itself is a relatively stressful business and that everyone sooner or later, by fair means or normal (to borrow a phrase from Lord Egremont), must come to terms with the fact that some people are more and some less intelligent than he is. Unless he does come to terms he is unlikely to lead a very satisfying life. One learns to cope with great stresses as an adult by dealing successfully with small ones in childhood and adolescence. The minor stresses of examinations may help in this process of adaptation. They may also help one to assess one's ability in a way which, despite all its faults, is considerably more objective than that of subjective personal assessment. There is perhaps only one method of testing ability which is less reliable than that of trial by examination. That is regular personal evaluation by one's teachers or by one's immediate superiors. University dons, for example, are a most heterogeneous collection of individuals. Some are completely reliable, honest and free of vindictiveness. Yet the academic community as a whole perhaps contains a higher proportion of apparently eccentric, unreliable, volatile and vindictive individuals than does any other profession. One does not have to spend much time in a university common room before realising this. I, for one, should not like to have my academic future determined entirely by those directly responsible for my education. I do not want to be committed to kow-towing to the don I do not like. I do not wish to avoid as though he were unclean the one I do like lest my fellow students should suspect me of seeking some unfair advantage over them. At least, when the examinees are identified only by number and not by name, the personality, status, rank and family of the unfortunate candidate does not influence the examiners' decision. This is one reason why I fail to understand the espousement of the cause of examination abolition by some of the more militant students. In the examination room, a man's unpleasantness, his smelly feet, his long and greasy hair, his soft fair skin and blue eyes are all irrelevant. Without an examination, the man with body odour may be doomed for ever by the personal opinion of a don with an over-sensitive nose. The nasty brilliant child who picks his ears at the back of the class may be condemned to obscurity by the hack conformity of a mediocre teacher who cannot cope with his intelligence.

The obvious fact can never be stressed too often and too hard.

The study of man by man has built-in fundamental sources of error which cannot even in theory be surmounted. The fact of making the measurement alters the behaviour of the man under study. This applies even to man's physical attributes and with considerably more force to his emotional and intellectual life. This is not to say that man should not be studied. In such studies attempts must always be made to reduce as far as possible the influence of the observer upon his victim, and that is why examinations are preferable to personal assessment. But we must never be deluded into believing that even such apparently objective assessment has any very solid reality.

Chapter 6

Science and Religion

Whatever the theologian or the trained scientist may think, the majority of people in the developed areas of the world seems to believe that science has destroyed the need for religion. Repeatedly one hears that God is now superfluous because science has taken his place. Fewer and fewer individuals, particularly among the young, feel that they must make a token religious gesture just in case. As science conquers ever-expanding territories there seems less need than ever for a God to fill the gaps in human knowledge. For the clever young intellectual God is often very, very dead. Many humanists have had to revise early over-confident views about the inevitability of man's progress, but few have needed to alter their opinion that God is finished. All enthusiasm for belief in him will gradually wither away as the mass of knowledge in libraries and computers becomes more impressive. God is an outworn dream buried beneath the intellectual brilliance of man. His headstone reads: "He served his time, but mechanisation has rendered him redundant."

The obituary notices are premature. Earlier chapters have demonstrated that the field in which scientific certainty operates is a very narrow one. It holds the classical physicist, the chemist, the engineer within its safe fences. Nuclear physics and biology are hopeful hangers on, but partly for practical and partly for theoretical reasons the view of nature which they provide is rather like that through the fog on a dark November night. One is never quite sure what is phantom and what is reality. But when man studies man the situation is very much worse even than that in biology and nuclear physics. No matter how much homage is paid to the scientific household god, reality and reliability can never be achieved. Whatever is examined changes, shifts, dissolves and melts away even as it is touched. The scientific study of man is a myth, perhaps the most dangerous of all the myths of modern civilisation. Ultimately the psychologist, the psychiatrist, the sociologist must each confess that his work must be prefaced by "I believe" and not by "I have proved scientifically". The intellectual basis for what the scientist says of man is no stronger than that for what the theologian says. By means of a gigantic confidence trick, by pretending that the study of man is science, by hanging on to the coat tails of solid, successful, reliable physics and engineering, an army of atheists and agnostics has forced many theologians to turn and flee.

What, then, does science have to say about God? Strictly speaking, nothing. The concept of God is a hypothesis which is untestable by the application of the scientific method. One cannot design an experiment to see whether God exists. The hypothesis is therefore scientifically useless and the existence of God cannot be scientifically proven. Equally and most emphatically, his existence cannot be disproved. Science can have nothing concrete to say about God because, quite rightly, only testable hypotheses are permitted their places in the scientific pantheon. The same, of course, applies to the problem of free will. It is quite impossible to design an experiment to see whether man really can decide his own course or whether everything is rigidly determined for him. Science certainly fails to demonstrate that man does have a genuine choice. Equally certainly it cannot demonstrate that the concept of free will is pious nonsense. Like all other approaches to the problems of God and free will, that of the scientist begins and ends with the creed, "I believe".

Although science has nothing concrete and direct to say about God, indirectly it provides pointers which some might find useful. In Chapter 2 it was shown that the absolute bedrock of science is a belief in the orderliness of the universe. The scientist believes that every event occurs in accordance with natural law. Events which appear to lie outside such law are merely occurring in accordance with laws which have not yet been discovered. The belief can never be proved true in any logically watertight sense, but the practical successes of physics, chemistry and engineering suggest that it is not unreasonable. There genuinely does seem to be order in the universe and within the limits set by the uncertainty principle the nature of this order seems accessible to investigation by the human mind. The ultimate aim of all pure research is to reveal this inevitable conformity which the scientist believes lies behind all the apparent chaos.

Why the universe should exist at all and, given its existence, why it should appear to be governed by immutable laws are questions beyond the realms of science. Hypotheses which attempt to answer these questions can never be objectively tested and are therefore scientifically valueless. However, many laymen and some scientists appear to be under the impression that science can throw some light on the matter. It is therefore interesting to see what happens when the scientific approach up to the stage of hypothesis formulation is crudely applied to the problem.

When confronted with such stupendous orderliness and law as the universe reveals, the first act of any intellectually alive man is surely to wonder why. I challenge the most cynical of human beings to stand in a Himalayan cleft on a clear night, to look up from the ice of Annapurna to the glittering stars, and to fail to wonder. Having realised that the problem exists, the next step is the formulation of a hypothesis to answer it. Innumerable hypotheses have been proposed, but essentially each can be put into one of only three fundamental groups. These are the agnostic, the atheistic and the theistic. The agnostic says that it is impossible for man to know the answer. This is a valid and eminently reasonable approach and is undoubtedly correct if by "know the answer" one means in a scientifically acceptable sense. But although reasonable, it is unsatisfying and sterile. The scientist who adopted an agnostic attitude when faced with the problems of cancer or of increasing the world's food supply would hardly get very far. This attitude takes one but a short way along a cul-de-sac. It leads to no mistakes but it also leads to no adventures and to no great discoveries. The second answer is to say that there is no Supreme Being, that God is dead, having been nothing but a reflection of man and a figment of man's imagination. This is the approach of the atheist. It represents a fantastic jump of faith. It does not say only the safe thing which is that we can never objectively prove the validity of any answer as to why the universe exists. It boldly asserts that there is no God behind the law. It is a difficult position to adopt. The atheist must either blindly assert that the order is simply there and that no explanation of its existence is required, or he must devise some answer which positively excludes God. The first bald statement is hardly very satisfying: it has all the disadvantages of the agnostic position without being sweetly reasonable. The second has never been successful and no worth-while atheistic explanations of the order exist.

The third answer is that God is a reality and that the order is a consequence of that. In many ways this hypothesis represents an approach most nearly like that of a scientist tackling some more mundane problem which, unlike this one, is genuinely within his sphere. Unlike the agnostic, the man who believes in God is at least prepared to stick his neck out and to try to find an answer. Unlike the atheistic attitude, a belief in God does account for natural law reasonably satisfactorily. It certainly does not require the colossal leap of faith needed to assert that there is no Supreme Being.

The only way in which a man can decide about the validity of the three approaches is to try each one for himself. The matter must be decided by the subjective beliefs of each man alone and the scientific method is of little help. There can be no question of proof. The agnostic approach is by far the most reasonable in theory and is undoubtedly correct if the only acceptable evidence is scientific. But I find it slightly odd that many agnostics who are prepared to take major family, financial or political decisions on the basis of evidence which is far from that required by science should be such sticklers for scientific propriety when it comes to the why of the universe. The agnostic's seems to me to be a dull, cowardly sort of attitude. I have much more sympathy with both the atheist and the believer. Indeed, the audacity of the man who says "There is no God" and boldly takes the consequences makes one gasp. I admire enormously the thoughtful atheist for, if he really understands what he believes, his stand is outstandingly courageous. I admire him even more if he holds firmly to his belief when faced with the reality of death. But though I admire the atheist because of his courage, I cannot help feeling that the evidence is just in favour of God.

Although I believe this to be a less important problem, much the same things may be said about miracles. They too are outside the realm of science and the scientific method cannot be used either to prove or to disprove their existence. Miracles may be defined as apparently unique events which seem unaccountable on the basis of natural law. They come in two varieties. The first type are not really miracles at all. They are events which, although apparently at variance with natural law at the time, are later seen to be in accordance with a law unknown to the observers of the miracle. An excellent non-religious and therefore, I hope, non-emotive example may be taken from Rider Haggard's thriller, *King Solomon's Mines*. The heroic explorers are held prisoner by a fierce African tribe and their fate is in the balance. It is clear to them that unless they can pull off some dramatic coup they are dead men. Then one notices in his almanack that an eclipse of the sun is due. The tribesmen cannot know this. The explorers inform the Chief that as a display of white power the sun will be darkened. The eclipse occurs, the Africans are suitably impressed and the immediate crisis is averted. The eclipse was obviously not a miracle at all. It appeared to be so only because the tribesmen were unaware of that aspect of natural law. The modern scientific tendency is to explain away all miracles on this basis.

The second variety of miracle is what might be called the true one. It is a unique event which genuinely cannot be accounted for on the basis of natural law, known or unknown. The believer says that such happenings reveal the hand of God. The non-believer asserts that science has demonstrated that such miracles cannot happen. They are merely figments of the imagination or at best happenings in accordance with as yet undiscovered laws. Science has demonstrated nothing of the kind. The scientist begins with the belief that the universe operates entirely according to law. He begins by believing that unique events which cannot be explained by natural law do not happen. He must believe this for if such unique events did occur on any large scale, the repeatable scientific experiment would be impossible. Since by definition, by act of faith, the scientist excludes miracles from the realm of science, he can hardly use science to demonstrate that they cannot occur. The non-occurrence of miracles is part of the scientific creed. It is therefore arguing in a circle to say that science demonstrates that miracles do not occur. The premise is the same as the conclusion. Even if the creed did not exclude miracles from the outset, the scientist could not use his method to prove or disprove the reality of a supposed miraculous occurrence. In order to be studied reliably by science, events must be repeatable. True miracles are unique and therefore cannot be made the subject of scientific investigation. I am not saying that true miracles do occur. All I am showing is that science has not demonstrated that they do not occur and nor will it ever be able to make such a demonstration.

As far as the existence of God and the occurrence of miracles are concerned, religion has therefore nothing to fear from science. If anything science actually tends to support the concept of a Supreme Being. What religion may have to fear is the uncertain light that science throws on the nature of that Being. Innumerable unpleasant consequences of the operation of natural law occur every day and there is no need here to detail examples. The scientist may quickly concede to the theologian the argument about the existence of God, but then he may go on to say "What sort of God is he who allows these things to happen?" It is the problem of the nature of God rather than that of his existence which provides the major stumbling block to the acceptance of creeds such as Christianity, where the central figure is a benevolent deity interested in the fate of every individual.

I think that a satisfactory answer is unlikely unless one divides the apparent evil in the world into two categories, that which appears to follow upon the operation of natural law independent of man's intervention and that which appears to be a consequence of man's own actions. The first category includes pain and death occurring as the result of natural events. The second includes the evils consequent upon wars, the development of industrial cities and all the varieties of man's exploitation by man.

To me, as a scientist, the first category gives little trouble. Some of the supposed difficulties about the nature of God which come into this group and which are continually being thrown at the theologians are not really difficulties at all. The first of these problems is that of death. Why should God allow the existence of such an apparently dreadful natural event? The biological answer is so simple that it may appear facile. As has been said of other phenomena, if death did not exist it would have to be invented. Without death this planet would rapidly be covered by an unimaginable mass of living things and would resemble nothing so much as a canvas by Hieronymus Bosch. Without death there could be no birth. Without death there could be no animal life because all animal life utimately depends upon food derived from the sacrifice of other organisms. Without death man would not exist because no evolution could have taken place. The mind concerned about the finality and cruelty of death may perhaps take some cold comfort from the thought that it is only by the courtesy of the gardener in white that it itself has been enabled to have life. The scientist need see no conflict between the occurrence of death and the existence of a personal God.

A second problem, that of pain, is on the surface a much more difficult one. It is the subject of many long and complex treatises by uncomprehending theologians and others. To the biologist the reason for the existence of pain is perfectly obvious and its advantages far outweigh its disadvantages. The ability to feel pain is perhaps the most important of all the senses. If I had to loose a sense, pain is the one I should fear the loss of most. I would much rather be blind than completely free from pain. This may appear masochistic but it is unadulterated self interest. Pain is vital if a young creature is to grow up normally and if an adult one is to pass through life relatively unscathed. Pain is the sensation that warns that fires are hot, that knives are sharp, that broken limbs must be rested, that medical attention must be sought. Without pain no wild thing could go beyond infancy without being crippled. Without pain even human infants with protective parents would be horribly deformed after only a few years of life. Without pain medicine would be a mockery, with few patients coming to the doctor in time to be saved from their potentially serious diseases.

Some readers may be sceptical of all this. There are very rare diseases in which the sensation of pain is absent from birth. I have seen only one small boy with this affliction. At the age of ten he had lost six fingers, his face was scarred and his limbs were twisted all because he received no warning when he was in danger of injuring himself. In other diseases the loss of pain sensation occurs later in life. Leprosy is the common example. The dreadful deformities whose photographs one sees in some missionary publications and in the advertisements for such organisations as OXFAM and War on Want may not be solely the direct results of leprosy itself. One form of leprosy, for example, attacks the nerve trunks and destroys the fibres which carry pain sensation. The deformities which then occur are the results of injuries which may be accidentally inflicted upon insensitive tissues.

The obvious counter to this explanation of the existence of pain is "Why should God have thought it fit to create circumstances in which animal life needs pain?" I do not think that this is a serious difficulty. An animal without pain sensation could move safely and without injury only if it were covered by a steely hard exterior which was completely resistant to damage by heat or mechanical trauma. In such circumstances pain would not be the only sensation missing. I for one would not like to do without the pleasures which come from the human possession of a soft and flexible exterior covering. The sensation of pain is a very small price to pay for the ability to move freely and to experience all the other sensations which are made possible only because we have bodies whose outer covering is soft and vulnerable to injury. Not every aspect of pain can be explained away entirely but the fact remains that pain is an overwhelmingly useful and important sensation and is one of the greatest blessings conferred upon the animal world. No theologian need feel that pain is one of the skeletons in the religious cupboard which must be decently hidden away before a man can be persuaded to believe. As I see it, as far as pain and death are concerned, there is no conflict between science and the concept of a personal God.

One's attitude to the evil which results from man's behaviour towards man depends largely on whether one believes that man has a will which allows him to act freely. If all man's actions are the inevitable consequence of the operation of natural law, and if God exists and is responsible for that law, then God must be a monster. He is then directly responsible for all the foul things which humans do to other human beings. The only proper attitude to such a Supreme Being would be one of impotent rage. While one was raging one might permit oneself a wry grin as one realised that the impotent rage was itself the inevitable result of outside circumstances and therefore meaningless. If, on the other hand, man's behaviour is not rigidly determined by his environment and by his history, and if he genuinely does possess freedom to act, then one can hardly blame God for all the evil in the world. Nor can one blame him for not intervening to stop such evil since the intervention would inevitably destroy the freedom.

The central issue of this interaction between science and religion is "Do the results of scientific research mean that every event in a man's life is rigidly determined by events entirely outside his control, or does science hold out the possibility that man does truly have freedom?" It can safely be said that there can be no certain answer either way. If one holds that everything man does is rigidly determined, one faces the awkward dilemma that this conclusion too is determined by outside events, and therefore that for man truth is a meaningless concept. It is logically impossible to build up a watertight case for rigid determinism because the determinism implies that all the steps in the argument are taken because of outside forces over which one has no control and which force one to a conclusion irrespective of its truth. Science, therefore, cannot deny the possibility of free will, even if only because as a consequence of this awkward dilemma it is impossible to design a valid experiment. Nor can science prove that free will exists. However, recent developments in both nuclear physics and in the computer sciences do suggest that the consequences of the operation of scientific law may not be so rigid as was once thought and that in the future even a machine may be able to make a real choice. Science is therefore not at all incompatible with the concept of a benevolent God who has by the operation of natural law brought into existence the species man which has genuine freedom to act and therefore to do good or evil to fellow human beings.

History of the Conflict between Science and Religion

Even if this possible compatibility between science and religion is comforting, even if the whole conflict has been a ghastly mistake, that makes the battle no less real. It does not make the rout of religion any less dramatic. It does not diminish the size of the victory which science has won. The prize was the allegiance of men's minds. No matter how spurious was the conflict in the eyes of thoughtful experts, to the majority of men the fight between these two great systems of thought was real. The majority of men, scientists and theologians included, seems to believe that science has won hands down. But why did science inflict such a crushing defeat? Why do so many people believe that God has been killed by a scientific experiment? These questions are still important. The defeated party in a battle does not recover its lost territory merely by pointing out that the issues over which the battle was fought are all irrelevant. If religion is to make any recovery it must understand the mistakes it made. If science is to continue successful it must avoid similar errors.

First of all, religion failed because it went to battle over trivial issues which were irrelevant to its main themes. These were issues susceptible to scientific investigation and with which religion was not competent to deal. Looking back, with all the advantages of retrospective insight, it is quite obvious that most of the positions which some theologians tried to defend were ludicrous and inherently indefensible. Yet, and this is an important point, they were not thought ludicrous at the time. Theologians and scientists alike thought them to be important issues. Each side believed that if it could win these battles it could win the war. If theologians themselves proclaimed the importance of the issues, laymen could hardly be blamed for turning from religion when it became obvious which way the

91

battle had gone. Few Christians today would consider that it is essential to believe that the world was created in 4004 BC, that the process of creation took just six days or that Eve was literally fashioned from Adam's rib. Yet these were all issues over which science and religion did battle. There can never have been much hope for the churchmen.

Thus, the first reason for the failure of religion was that the theologians had not thought out their own positions rigorously. Thoughtlessly they were led into defending fantastic points of view which had never before been seriously challenged. Because they failed to change from within they were defeated by change imposed from outside. The second reason for the disaster is closely related to the first and again shows culpable idiocy on the part of those who defended religion. They had not thought out the consequences of their own arguments and they took up what has been called the "God in the gaps" view of the world. "We admit that the scientific method is a powerful tool for investigating the nature of the universe. We realise that it has been successful in many ways but, in spite of this success, science cannot explain this and this and this. These are important gaps in scientific knowledge. They cannot be filled by man and God must be invoked to account for the existence of these mysteries." But the gaps have become smaller. Repeatedly science has produced mundane, natural explanations for what seemed to be mysterious events. Repeatedly religion has been forced to retreat and retreat and retreat, desperately seeking new gaps to sustain the myth. On innumerable occasions laymen have seen science fill in an awkward gap whose strange features were once thought to require God for their explanation. The inevitable result has been a shrinking God.

Some religious thinkers have still not learned their lesson. For example, in November 1968 there was a controversy in the correspondence column of the *Guardian* about the problem of God's existence. A point which some thought to be very telling was that science could not explain extra-sensory perception. One writer imagined that God must exist in order to account for such unusual methods of communication. This is an excellent example of the God in the gaps argument. What should have been said is that science cannot explain extra-sensory perception *yet*. There seems every reason to suspect that what is now wrapped in obscurity will in the future be accounted for on the basis of as yet undiscovered scientific laws. Where, then, will the *Guardian* correspondent's God be found? It is such a pity that she fails to understand that God is more necessary to explain the existence of law than he is to account for the gaps where such law does not yet seem to apply.

Because religion tried to defend what was essentially indefensible and because it adopted the God in the gaps form of argument, it came into conflict with science in a totally unnecessary way. Many churchmen genuinely thought that science was out to destroy true religion. Because of the stupidities of the religious apologists, many scientists thought that there really were issues over which science and religion should fight. Based on a false understanding of both science and religion a real and furious war developed. The scientists won all along the line and the Christian God seemed to be a shrinking idiot. The battles over the date of the creation and over the time taken to make the world were over almost before they began. But some scientists forgot the vital maxim that battles are not wars. Because science was in conflict with some current religious teaching they tended to think that it was in conflict with all religion. Because science had destroyed some cherished aspects of religious belief, they imagined that science could destroy all religious belief. The layman spectator watching the struggle and seeing religion routed in the great public controversies cannot be blamed for failing to realise that there was no real ground for conflict. Religion was finished: why not turn to hopeful, successful science as the new saviour?

Making a more important contribution to the ascendancy of science than all the academic arguments was the fact that science was apparently so overwhelmingly powerful in practical terms. From the time of the Roman Empire until the early eighteenth century, religion had been the dominating force in men's thought, and yet man's command over his environment had increased hardly at all. Technologists in the early eighteenth century could do very little that was not done equally well by the Romans. And then science-based technology began to exert its influence, and within decades the face of Europe was transformed. Science seemed to be capable of so much, religion of so little. Who could fail to believe in this new deity which had so much more power over people's lives than the old, tired, outworn God? Again one gets this dangerous extrapolation of thought. Science has defeated religion here and here and here: it must be capable of defeating religion everywhere. The argument is false and the conclusion over-confident and dangerous. But both the argument and the conclusion are very appealing to the ordinary man who above all wants to be associated with success and not with failure.

In summary, therefore, science won because the theologians regarded as vital, ludicrous beliefs which were totally irrelevant to religion's main theme. The scientists had no difficulty in demonstrating the falsity of these peripheral issues and both they and the laymen spectators thought that they had undermined the fundamentals upon which religious thought is based. Secondly, the theologians completely failed to understand science. They knew too little about it to expose its underlying assumptions and fatal flaws. They employed the gaps argument which inevitably led to a shrinking, weakening God. Thirdly science came out on top because it seemed to be limitlessly successful. The very narrow range within which science is a truly valid approach had not yet been pointed out. The significance of the uncertainty principles was unknown. It was not realised that all statements which scientists make about man are of questionable reliability. As a result science annihilated religion in all the great showpiece intellectual confrontations. It would be surprising if laymen had failed to get behind what they thought were the big battalions.

Lessons for Science

In the mid-nineteenth century, religion was vulnerable because few theologians had bothered to think out rigorously what should be their attitude to the modern world. As a result, Christianity fought and lost an unnecessary battle. If theologians themselves had put their own house in order, religion might well have come through into the twentieth century virtually un- scathed. But by taking up unwise and exposed positions, the churchmen forced the scientists to attack them. When these positions fell, most men imagined that science had replaced religion. I have a suspicion that science itself may now be making mistakes almost identical to those made by nineteenth century Christianity. I wonder whether the swing away from science in the schools may not be the cloud no bigger than a man's hand which heralds an onslaught on science comparable in ferocity to that which destroyed nineteenth-century religion. If the churchmen can understand and appreciate what is happen- ing, religion may be able to stage a dramatic comeback. Un- fortunately, few clergy show any signs of really getting to grips with science and I think that their opportunity will be wasted. I only hope that nothing more sinister arrives to take advantage of the vacuum in thought.

First of all, in a manner of which any unthinking nineteenth- century bishop would have approved, many scientists are defending with untoward vigour positions which seem to me and probably to most people to be untenable. The most extra- ordinary of these positions is the statement that the scientist has the right for curiosity's sake alone to investigate anything and everything and that he cannot be held responsible for the uses to which his research is put. There is an element of real truth in this and that is why it is so attractive and so dangerous for science. If the concept is pushed to an untenable extreme, when that extreme position is destroyed, as it surely will be, the truth will be lost with the idiocy. The truth is that, particularly with pure research, it is often quite impossible to see what the end will be as the stories of Hertz and Faraday demonstrate. There- fore it is unfair to make the scientist responsible for outcomes which he could never have conceived. Yet even this impossibility of knowing all the consequences is not a complete defence. There is always a definite risk that the outcome of research may be very bad. Since this risk is there, and since it cannot be catered for, some might say that research should be done only if we can be very nearly certain that the end will be good. Of course, if that condition were imposed virtually no research at all would be carried out. To the modern world, conditioned to the belief that change and progress and the good of mankind are all synonymous expressions, this would seem a disaster. And yet I sometimes wonder if it really would be such an appalling thing.

For example, consider medical research, the sort with which I am involved and which is universally regarded as a worthwhile activity. It is arguable that medical research is the most destructive and least controllable weapon ever let loose upon mankind. Before modern medicine and public health arrived on the scene, most societies had achieved some form of equilibrium. Birth rates and death rates balanced out, each man could know that his skills would not be rendered redundant by rapid change, and there were no threats other than those of war and disease to which man had become adapted over thousands of years. I am not suggesting that in those societies the lot of the individual was idyllic. Especially in terms of personal comfort, it quite obviously was not. But there was a stability, a sense of being part of the cycle of life which hardly exists today except in isolated rural communities. Too, because of the relatively high death rate, young men in any field had a reasonable chance of achieving real responsibility at an early age without waiting over long for the shoes of the departed. They were thus less likely to become frustrated and disillusioned. And then came modern medicine and public health. The death rate fell precipitously and the population rose correspondingly. In Europe, although the condition of the industrial poor was miserable, the population explosion did not lead to disaster. Agriculture advanced with medicine and managed to keep pace with the food needs of the people. Industry was at a stage when mechanisation was primitive and enormous numbers of people were required to man the great new factories. Those not absorbed in this way could always emigrate to the new developments of America, of South Africa or of Australia. Our ability to feed and to employ people was therefore not hopelessly outstripped by the falling death rate. We seem to think that this can happen again in Africa, in Asia and in South America. But we are living in a fool's paradise. We have reaped all the advantages of modern medicine and have escaped most of the disadvantages. But we may be handing on to less fortunate peoples a terrible legacy, a true kiss of death.

Medicine, for the underdeveloped countries, is relatively cheap. It is also emotionally attractive and draws many dedicated souls and large sums of conscience money. The establishment of industries to give employment and of advanced agricultural methods to supply food are not so emotionally attractive and draw much less support. Even those industries which are developed tend to be highly mechanised and to employ relatively few well-paid individuals. The masses of young people now growing up, given life by our medical aid, have no work to do and no food to eat. They are too numerous to be accommodated within the framework of traditional society, and that society has been shattered. Especially dangerous is the massive unemployment amongst the relatively educated who gaze with hungry eyes at the fortunate few who receive what are comparatively enormous salaries. No wonder that the men with power and influence feel that they must hang on whatever the cost or they will go to the wall. Unless we do something about the balance between medicine, on the one hand, and agriculture and technology on the other, the situation will become impossible to control. The next hundred years will then see starvation, inhumanity and war on a scale which dwarfs anything that has happened before. Is that what the believers in medical research want? Is it not conceivable that had they not opened Pandora's box the state of the world might have been better in fifty years' time than it is going to be? I do not know, but the matter is at least arguable.

Unfortunately the experience of those who have tried to keep agriculture and technology advancing at the same rate as medicine has not been happy. In theory it is the right answer, but in practice it does not seem to work. This is mainly because the medical measures required to bring about a dramatic reduction in the death rate are simple and cheap. In contrast, the development of advanced agriculture and technology is complex and expensive and requires highly trained people. In any case, even if agriculture and technology do advance, there must be a theoretical upper limit to the amount of food that can be produced on this planet. In contrast, short of starvation or war, there seems no reason why the population should not go on expanding indefinitely. Research into industrial food production can only postpone the disaster, it cannot prevent it happening. This means that the only real hope is for medicine to devote itself as energetically to restricting birth as it has in the past to defying death. Only in this way can a reasonable, permanent population balance be achieved.

With most types of applied research, the scientist can have no excuses. Many of the possible end results are perfectly clear from the outset. There can be no avoiding the conclusion that any scientist who carries out such research must bear a large part of the responsibility for the consequences. At the beginning, only he can know enough about the matter to say what the likely implications are. It is therefore his duty to explain to the public as fully as possible and as early as possible what the likely end results will be. The popularisation of science instead of being sneered at by "proper" research workers, as it often is, should be part of the responsibility of every scientist. If the likely outcome of a line of research could have untoward consequences for society, then it is the scientist's duty to withdraw from that research as soon as possible.

Weapons development is an obvious and much-discussed problem where the issues are by no means so clear cut as they are made out to be. The scientist who works upon it can legitimately say that, although he does not like such research, he feels that he must do it in order to ensure the survival of the human race in general and his own society in particular. The issue is an intellectual and not a moral one. Two scientists may be equally concerned about the future of humanity. One may decide that, repugnant though it may be, the balance of terror is the only way to ensure survival. The other may feel that the manufacture of weapons of mass destruction, whether nuclear, chemical or biological, increases rather than reduces the risk of world catastrophe. The two hypotheses are quite untestable and there is no way of saying which, if either, is correct. What is certain is that neither one is more moral than the other. Both scientists are devoted to the same highly moral end: they merely differ about the practicability and reliability of the means to that end. There is nothing more nauseating than the claim by those who are against weapons research that the scientists who carry out such work are a lower form of moral life. Most of the people I have met who are working on military hardware genuinely believe that what they are doing is the only way to ensure freedom from a major world war. They do not mind other scientists being of the opinion that this belief is incorrect. But they get very upset, and quite rightly so, when other scientists and laymen say that this belief is immoral. We are likely to get a sane weapons policy only when the emotive moral heat is taken out of the argument.

But there are other areas of science in which the issue of whether or not a piece of research should be done does seem to me to be a genuine moral one. In these fields there is a definite risk that research, while conferring no great advantage upon humanity, is in clear danger of doing great damage. For example, I think that one line of research which should be stopped forthwith is that into methods whereby human parents may choose the sex of their children. Unlike many of the other biological stories published in deliberately sensational books, this is undoubtedly something which is going to be possible within a very few years. Yet I submit that this is knowledge which by deliberate choice we should refrain from seeking. The adverse effects of such a choice exercised on a large scale could be socially disastrous. Parents in most societies seem to prefer sons to daughters. This could lead to a gross excess of males and to all the horrors and violence of a society in which this situation occurs. The research should be stopped at once. The scientists know what they are doing. They will be responsible for the consequences.

On Teeside in northern England in early 1968, over twenty babies died in an epidemic of gastroenteritis. The germ which caused the disease was resistant to most antibiotics. It may well have been produced as the result of a quite reckless application of science to agriculture. It is becoming more and more common to fill young animals with foods containing small quantities of antibiotics. The drugs reduce the number of bacteria in the gut and enable the young animal to grow more rapidly. It is also common to use antibiotics to treat mastitis in one infected udder of a lactating cow. The antibiotic goes into the milk from the other uninfected udder and into the supply which is sold in the shops. Nothing sinister about all this, one might think. Unfortunately, it is well known that the small doses of antibiotics used kill some but not all of the germs. The ones which survive and their descendants may be immune to the effect of the drug. They may transfer this immunity to other types of bacteria with which they come into contact. Infections due to such germs cannot be treated properly and those who suffer from them cannot easily be saved. Until now the pharmaceutical industry has managed to keep just ahead by developing more and more new antibiotics. But there is no guarantee that this success will continue indefinitely. There is a serious danger that by the over-use of antibiotics in agriculture we shall produce a host of virulent organisms over which we have no control. In order to gain a transient commercial advantage, agricultural scientists will have helped to squander one of the greatest healers given to man.

These are just two examples where some scientists who know what they are doing are acting in a way which seems to me indefensible in order to obtain a short term gain. If scientists continue to defend the concept that research workers are not responsible for the uses to which their findings are put, then science will become more and more disreputable in the public eye. Since science is largely financed by public money, this can mean nothing but impending doom for science. The adoption of an extreme position will ensure that babies and bath water go slithering down the drain together.

On the other hand, it is important that scientists should not refrain from defending themselves vigorously and publicly when the blame is unfairly laid upon them. The decision to work towards the atomic bomb was a political one. Many pure scientists who in peace time would have abhorred applied research of any form, let alone research on weapons, devoted their energies to it because they believed that for the survival of a moderately humane world it was essential to get there before the Nazis did. In fact, of course, the bomb came too late to be used against Germany. The scientists are often blamed for the tragedies of Hiroshima and Nagasaki, and without doubt it was they who provided the means for the destruction of these cities. Yet it was the men who developed the bomb who pleaded that its power should be demonstrated by first dropping it on some deserted enemy area and not on a city. The Japanese might or might not have been impressed but at least nothing would have been lost had the experiment been tried. The decision to destroy the cities at the beginning was political, not scientific. Perhaps the lesson to be learned from this is that some men will always misuse the power which science puts into their hands. Since once the discovery is made the scientist can do nothing to prevent its misuse, he should perhaps think harder at an earlier stage.

The assertion that the scientist is not responsible is one example of an extreme position which, once destroyed, may bring down with it many more worth-while things. Another such extreme belief is the thesis that science can and should be applied to anything and everything. In some ways this is the converse of God in the gaps theology. Oddly enough it is the most successful scientists in the physico-chemical field who are least likely to think in this way. It is those who misleadingly call themselves social and political scientists who, without any real understanding of what science is about, are most active in disseminating this point of view. In brief, the argument goes that because science has been able to tackle successfully many complex problems, it is capable of tackling all problems given time and a big enough computer. I have already demonstrated that this is quite untrue. The uncertainty principles and the need for testable hypotheses and repeatable experiments mean that the range of reliable science is strictly limited. The problem is not that there are too many variables nor that the amount of information which must be collected is too vast to contemplate. These are practical problems which in theory, though perhaps not in practice, could be dealt with by a computer. The problem is the nature of the scientific thought process itself. Attempts by certain apologists to over-extend its validity can only lead to a disaster for science comparable to that which the God in the gaps concept brought upon religion.

Again, like nineteenth-century religion, twentieth-century science is in danger of decline because it is obvious that it has failed to provide answers to any of the really important problems of the world. In spite of the phenomenal growth of scientific activity, if anything the problems are becoming more serious. What is worse, science is actually responsible for some of the most unpleasant disasters. World population is exploding, our cities are clogged with fumes, our countryside is despoiled, our planet is being raped of its irreplaceable natural resources, and science is quite rightly being tarred with the brush of vandalism and failure. The same accusing mind which says, "What's the point of God if he allows such things to happen?" may simply substitute science for God with equally devastating results. Science will survive only if it controls itself, only if it takes care to stress the limited areas over which it is valid, only if, no matter what the pressures, it refuses to go beyond its proper frontiers, only if it ceases to proclaim itself the panacea for all ills. If it does not do these things a new accuser with the devastating power of a T. H. Huxley may arise. This time the target will be science and not Christianity. Dr. Leach, Provost of King's, may be made to look as foolish as Bishop Samuel Wilberforce.

And finally, science is losing its power over the mind of man and in particular over the mind of youth because the careers of scientist and engineer do not lead to the gods of great wealth, great fame, great glamour and great influence. Who remembers the names of the Nobel prize winners for more than thirty seconds after the announcement of those names on the television news? Which great scientist has a face which is instantly recognised when it is flashed on to the television screen? Which scientist by reason of his science has built up an enormous fortune? Which scientist has any influence on the decisions of government? The scientist must be satisfied with £2,000 a year, no glamour, no fame and no power and, if he is lucky, the satisfaction of realising that he has added a few sand grains to the Sahara of human knowledge. Is it so very puzzling that the bright young things are not anxious to go into the science sixth?

Chapter 7

Science and Education

The relationship between science and education has at least three major aspects: the role which science should play in the education of every child, the nature of the educational process by which a scientist is made and the light which science throws upon educational theory. The first two will be discussed in this chapter and the third in the next one.

If the reader is to understand why I feel that science has such an important part to play in all aspects of education and why I feel that at present it is often being badly misused he must have some idea as to what I believe about education in general. I look at it from the viewpoint of a teacher as well as a research scientist and as I see it the ultimate aim of education is to achieve successfully the transition from totally dependent infant to independent adult. Education must give the young person experience in the use of his own mind. It must give him confidence in that experience so that he is able to consider and to make the right decisions about his life without continually depending upon others. In this chapter this training of the mind will be termed general education. Clearly it should be given to every child, irrespective of what his or her destiny is to be when formal education is over. In addition, education must also provide some form of specific vocational training which will enable the young person to earn his own living and to become financially independent of both society and his own parents. These two aims, the general training of the mind and the specific vocational training directed to a particular career, are quite distinct and must be kept clearly in view if any sense is to be made of a discussion on education. All men and women require both if they are to fill satisfactorily their places in the world.

General Education

The general training of the mind must come first. This training has essentially three components. The child must learn about words, must learn about numbers and must be encouraged to realise that he is different from all other children and is capable of thinking for himself. Words and numbers are the twin foundations upon which all intellectual activity is based. The child who does not early acquire facility in their use is handicapped for life. It is essential that children should be allowed to use their own minds in their own ways from the very first days. First, the child must be encouraged to express subjectively his reactions to himself, to his family and to the world about him. This can be done by several methods, for example by writing and painting. Secondly, the child's natural curiosity about the earth must be stimulated and guided. And there is no better medium for this than science. Every child is naturally a scientist. Every child wants to know why and how and where. Science, far from being an unsuitable subject for small children, is perhaps one of the most natural. Experience in the United States and with the Nuffield Project in England has helped to illustrate how eager children are to ask questions and to design simple experiments to answer them. In this way the child's individuality and questioning can be encouraged. At the same time the young mind can be introduced to the concept of objectivity and of the unbiased assessment of a situation.

Facts and Principles

In general education, there is no place for the cramming of the child's mind with a multitude of facts. The learning by rote of vast amounts of factual information is an utter waste of time at any level of non-vocational training. Such learning stultifies the mind and prevents the child from realising that his brain is a delicate and flexible instrument which can be used to tackle problems more related to real life than the trick mastery of railway time tables and telephone directory pages. It prevents the young person from using his own initiative and insidiously encourages the belief that if the answer to a problem has not been memorised the problem cannot be tackled. Yet this fact-cramming dominates British education up to the age of 18, American education between the ages of 17 and 24 and African and Asian education at all ages. Some salvation comes to the Europeans and North Americans because in the final stages of university training they are at least encouraged to use their own minds. But we seem to have handed on to many African and Asian countries the worst features of our educational system without any of its saving graces. There is nothing more terrifying than being confronted by African university students who, with admirable diligence, insist on learning word for word what the textbook says but who find immense difficulty in thinking about those words themselves and hence in applying in unfamiliar situations the material they have learned. It is not their fault. It is ours. We have all too often sent to them teachers whose knowledge is so inferior that they too have felt insecure when going outside the written text. The only way in which such teachers can be certain of being right is to insist on the forms of words used in the book. The tragedy of such rigidity is all the more appalling because it occurs in parts of the world where the major characteristics required of an educated person are flexibility and adaptability.

Once the use of words and numbers have been mastered, and if facts are not to be learned by rote, what then are the qualities which general education should aim to develop? The first is an understanding of where and how to find facts when necessary. The young person should be made to understand that it is a waste of time to clutter his brain with a vast collection of information when with relatively little effort he can obtain the facts he requires from other sources. Children should be set simple problems of research which will teach them how to use libraries and to make files on subjects in which they are interested. An easy acquaintance with these simple techniques of information retrieval will later be much more useful than a magpie memory.

A second aim, closely related to the first, is the development of the ability to marshall facts to demonstrate a principle or to prove a point. This can be done by changing the nature of the project. When methods of using a library are first being taught, the questions asked should be purely factual, such as, "How many planets are there and what are their names?" or, "What important events happened in the year 1542?" or, "What are the major rivers and mountain ranges of the Indian sub-continent?" But later on, more controversial issues may be raised which require not only the collection of facts but also their selection and marshalling. Science, history, religion, economics, geography and politics are all fertile fields from which such problems may be drawn.

The third aim of general education, and one which is perhaps the most important of all, is the establishment of a deep under-standing of the fallibility of the human mind. Every form of human intellectual endeavour has certain features in common. There must first be the recognition that a problem exists, then relevant facts must be collected and finally opinions and hypo-theses must be formulated on the basis of these facts. Every young person must come to understand that up to this point human intellectual processes are inherently unreliable and that there is usually no such thing as the right answer to a problem. Five equally clever men may have access to precisely the same information and yet may express five different opinions about a particular issue. Their answers depend more on their precon-ceived ideas than on the facts available. Only by introducing this concept to children at an early age can they be provided with adequate defences against dogmatic and intolerant opinion.

The Place of Science in General Education

Science has a place in general education because it offers one way by which greater reliability can be introduced into intellectual activity. Some would go so far as to say that it is the only way to ensure that man's mental processes come to conclusions which are at least partially valid. But if science is to have a place in general education, more time should be spent on teaching the fundamentals of the scientific approach than on teaching the facts about individual sciences. The facts will almost certainly be forgotten quite quickly once formal educa-tion is over, whereas an understanding of what makes the scientific approach different is likely to remain. It is particularly important for young people to understand how extraordinarily

106

limited are the fields within which the scientific method can be expected to produce valid results. They must realise that scientific reliability applies only in the very narrow fields in which controlled and repeatable experiments are possible. They must be taught to be on their guard against the use of the adjective "scientific" in contexts where it is inapplicable. If they are not aware of this they may frequently be misled into believing that many subjects where the scientific approach cannot be used are as reliable as classical physics and engineering. It is particularly important to appreciate that in the study of most aspects of man's mind and of his political and social organisation, reliable experiments cannot be carried out. The names "social science" and "political science" are nonsense expressions. They usually refer to the gathering of many facts and to the formulation of hypotheses, but not to the experimental testing which is the centre of science. A subject is not made scientific because a lot of facts have been gathered. It is made scientific only by subjecting hypotheses to controlled and reliable experimental tests. This point will be discussed more fully in the next chapter but it is mentioned here because it is the most important aspect of science to be taught at the level of general education. Young people must understand that it is simple dishonesty to call such subjects sciences in the hope that the unwary and the the uninitiated may grant them a status which they do not deserve. It is in the scientists' own interest to insist on a proper use of the word science. If they do not protest loudly against its prostitution they risk being blamed for the foolish actions of many academics who proclaim that their subject is a science without any real understanding of what science is about.

Vocational Training

The aims of vocational training should be quite different from those of general education. Much discussion of educational problems has been confused because the two have not been adequately differentiated. Vocational training implies education for a specific job. It should come only after a general education in which the young person's mind has been trained to tackle problems of many sorts. In vocational training two things come to the fore which have little place in general education. First of all, some facts are important in themselves. They must be memorised. The doctor called in emergency to the scene of a train crash must know as a fact the safe, pain-killing dose of morphine. The surgeon carrying out an operation must know as a fact the details of the normal and abnormal anatomy of the part with which he is dealing. The engineer must know similar facts about the machinery he uses, the businessman similar facts about the product he is manufacturing or trying to sell, and so on.

Secondly, vocational training must provide practical experience in real situations. It must gradually introduce the student to the job, so that by the end of the course he is capable of doing it entirely by himself. This is why, for example, it is laid down that every doctor after passing his examinations must spend at least one year in hospital working under supervision before he is allowed to go out and practice on his own. This is the principle on which many modern industrial and business training schemes are based. Practical experience on the job is intermingled with courses on the theoretical background. It is an essential feature of any worth-while course of vocational guidance that theory and practical experience must be mixed together as intimately as possible.

Importance of Specialisation

Recently, discussion about education in Britain has been almost hysterically preoccupied with the dangers of narrow specialisation. Three important government-sponsored investigations all emphasised this. The Dainton Committee[6], which studied the declining popularity of science in British schools and universities, and the Swann Committee[7] and the McCarthy Report,[8] which considered the employment of scientists, all united in a massive attack on specialisation in education. The argument, crudely put, goes something like this. "Britain is in a mess. America is strong. American education does not allow a concentration on only one or two closely related subjects until after the first degree course has been taken. The British system encourages such specialisation from the age of 15. In order to correct what is wrong with Britain we must copy the Americans and stop this specialisation." The fact that there may be other differences between the two nations which account for the different degrees of economic health hardly seems to enter the educationalists' passionate heads. And so we all must incant together: "The specialist becomes fossilised. The specialist is soon out of date. Only the non-specialist retains an open mind. Only the non-specialist can keep up to date with modern knowledge. The non-specialist will be our saviour. Long may he reign."

As yet neither those who attack specialisation, nor those who defend it, have shown much understanding of its real virtues. "Fossilised discipline", says the progressive anti-specialist. "Soft option", replies the stern advocate of specialisation. Debate conducted on this level is unlikely to produce worthwhile results. But specialisation does have real virtues both in the realm of general education and in that of vocational training. It is a pity that these virtues should be ignored.

First consider specialisation in general education. As I suggested earlier, ideally general education should lead the mind of a child through various stages until that mind is capable of collecting evidence to prove a point. Obviously this training must begin by considering problems which have been studied many times before. The child must be shown how these problems have been tackled and must be encouraged to find answers to them himself. But the whole business is something of an artificial game until the student reaches what are conventionally called the frontiers of knowledge, and has to tackle problems whose answers are not to be found in the back of the book or in the teacher's reference copy. No retrospective analysis of the way in which a problem was solved can ever really make a student understand what it was like to be faced with that problem before the answer was known. If a student is to understand properly the processes by which problems are solved and tackled he must be faced with intellectual difficulties to which no answer is yet known. Only the exceptional student will be able to provide original solutions but almost all can learn to appreciate the difficulties.

And therein lies the virtue of specialisation during the general training of the mind. If a student works at many subjects he can quite clearly know only a few facts about each one and can understand each one on only a superficial level. This is not to say that a non-specialist education is a soft option or that it lacks discipline or rigour. It is often a very hard option because the student has to know an enormous number of facts about many subjects. He has to work extremely hard in order to cover the necessary material. But despite its rigour and difficulty, the non-specialist education is dangerously defective. Because there is not time to study all or even most of the facts referring to a particular topic, the non-specialist has little chance to form his own opinion about that topic. He must accept the facts presented to him by his lecturers or by his textbook, and he has little time to question whether they are the only facts or even the most significant ones. Even if he makes up his own mind on the basis of the facts he receives, those facts are so highly selected by someone else that he cannot genuinely form his own opinion of a situation. He might have drawn quite different conclusions had the complete range of information been available to him. His ideas therefore depend entirely on slanted information provided by others. Furthermore, because only a few facts are pre-selected for the student's consumption, because he has time to read or to listen to only one opinion about a topic, the non-specialist can never appreciate the appalling complexity of real problems. It is the non-specialist who sees the world in wicked black and virtuous white and who is the source of all extremism. The specialist who has studied any single topic to its limits realises the difficulties and understands that almost all problems are morally grey. He is much less likely to express extreme and foolish views.

Another virtue in general education of a narrow concentration on only two or three subjects is that the specialist has very much more spare time than the non-specialist, whose every minute is occupied with the accumulation of mere facts. The non-specialist spends so much time rushing from one course to the next that he has no time to sit and think, to develop his own interests and to come to his own conclusions. When formal education is finished in this rushing world, there is little enough time for thought and for the development of intellectual interests outside one's work. It seems a pity that we must carry over this headlong rush into the educational system and effectively prevent most students from having enough time to think for themselves.

In summary, the purpose of specialisation in general education is not to learn more and more facts about less and less, but to acquire an understanding of the complexity of real problems and of the methods by which they are solved. The method, once understood, can be applied to any problem and the specialist thus acquires confidence in the use of his own mind. The non-specialist, who has superficially studied all problems and who has received his opinions in pre-packaged containers, has not fully understood any problem. He has driven himself mad by trying to fill his bursting brain with facts but, despite this, he has not acquired the intellectual techniques which will give him the confidence to use that brain. He has made himself into a good piece of blotting paper, but that will help him little when his teachers and his books are not there to guide him and he has to use his own mind himself.

In vocational training, specialisation is also important but it has quite different functions. Here, as its detractors claim, it is a narrowing, a concentration upon acquiring more and more facts and greater and greater skill in a smaller and smaller field. Put that way it sounds dreadful. But if someone in Professor Swann's family is unfortunate enough to require a serious operation, I wonder whether he will ask the local general practitioner to perform the operation in his surgery? Or when a complex piece of electronic equipment in Professor Dainton's laboratory goes wrong, I wonder whether he will call in the local handyman who fixes radio sets? I also wonder whether Mr. McCarthy, when he wins £300,000 on the pools, will allow the investment to be handled by the counter clerk at his local bank? The point is obvious. Most vocational tasks are complex and require a great deal of skill. Only the specialist can acquire that skill and can devote enough time to the subject to be fully conversant with it. The man with the highly specialised training is soon out of date and soon becomes an old fossil, the hounds glibly cry. Which is odd, because the non-specialist can never devote enough attention to anything to be up to date in any topic. If the non-specialist is never up to date, it is hardly fair to criticise specialist training because it soon becomes out of date. When it comes to obtaining hard advice about something which really matters to us, it is the specialist we all want, not the general handyman. Because the specialist knows what it is like to be up to date, he is much more likely to make an effort to keep himself there by continual re-education and training, than is the non-specialist who has never known what it is like to master completely the intricacies and complexities of any single field.

Science Education at School

Clearly all education must begin by teaching children about the use of words and numbers, but I do not think that this is the place to attempt to make detailed suggestions about education before the age of 14 or 15. The only thing I do believe is that science is an interesting subject for children below this age and that it should play a much larger part in most curricula than it does today.

At about the age of 15, those children who stay at school and who wish to continue with general education should specialise. They should specialise not in order to learn a lot of facts which will help them in some future career, but in order to gain some insight into the way in which the human mind works and into the methods whereby problems may be solved. However, I seriously doubt whether the form of specialisation now employed by British schools is the right one. For a long time there has been a tendency in many schools for British school children, between the ages of 15 and 18, to study only two or three subjects. Almost invariably these subjects come either from the sciences or from the arts, but not from both. Thus, at the age of 14 or 15, a young person effectively has to decide whether he wants to become a scientist or not. He may study some combination of mathematics, physics, chemistry and biology on the one hand or, on the other, two or three subjects chosen from English, history, geography, economics and the foreign languages. In this I agree with those who are against specialisation: no one should at this age be forced to choose between arts and science. But I do not think that the answer is to make the boy or girl study a multitude of topics. I would suggest that only three subjects should be studied and that ample time should be left spare so that any other topics which arouse interest can be followed up. First of all, everyone should study to a reasonably high level an important living foreign language. More than ever before, future generations of young people are going to meet foreigners, and even though English is rapidly becoming the lingua franca of the modern world, for the sake of courtesy if for nothing else, we too should make the attempt to understand someone else's language and culture. The second subject should be one in which scientific experiment is possible. It would matter relatively little whether this were a physical, chemical or biological subject, since the aim would be to teach the principles which are basic to all science and not the facts of any particular science. The real power, but strict limitations, of the scientific approach should be understood by every modern person who aspires to be called educated. The third subject would be one in which scientific experiment is impossible. It might be history, English, politics, economics, sociology or any one of a host of others, depending on the resources of the school. The aim of teaching this would be to illustrate the complexities and uncertainties of controversies which cannot be settled by the use of the scientific method. By following such a course, every intelligent school child would learn a foreign language and would be equipped to understand the principles behind the two great types of intellectual approach at present current in the world.

Science at the University Level

At about the age of 18, every child who wished to could take an examination in the three subjects he had studied, and on the basis of this examination it would be decided whether to admit him to a university. Although some general courses should be provided, I believe that in the main most students at this age know whether they want to read a scientific or a non-scientific subject and which one they want to do. I do not think that the age of 18 is too early to take this responsible decision. It is by denying responsibility, not by giving too much of it, that we have got ourselves into the present inter-generation dog fight. In any case, since the first university degree is primarily aimed to be a training of the mind rather than a preparation for a specific career, relatively few jobs would be excluded by a wrong decision at this age. If one chose an arts subject it might be difficult later to become a doctor or an engineer or a research scientist. But even if the wrong decision is made, it is not irrevocable. I know at least half a dozen doctors who read arts subjects for their first degree and I know one nuclear physicist who started his university life as a classicist. In fact, since I believe that the first degree should be a training of the mind, it seems to me that the nature of the subject studied is almost irrelevant so long as the course fulfils the following conditions. It should interest the student, it should force him to think about unsolved problems and it should leave plenty of time for extra-curricular activities, intellectual or otherwise.

I am not suggesting that because a student does, say, physics and politics at school he would have to do physics or politics at university. If someone has done well in physics at school he is likely to be able to do well in physics or in any other scientific subject at university. The same goes for politics and non-scientific subjects, and the new undergraduate should be able to choose freely what he wants to study, irrespective of what he has done in the past. All that would be required is a relatively minor reorganisation of university time-tables. Those who had already studied at school the subject which they wished to read at university would do a three-year degree course. Those who wished to read a different subject would spend the first year in a preliminary course covering the ground which others had done at school and would take their degree after four years at university in all.

The nature of the science courses at most universities leaves much to be desired. At the moment it usually consists of compulsory lectures and practical classes, all day every day from Monday morning to Saturday lunch time. The odd seminar may occasionally be thrown in as a titillatory afterthought. Personally, I find this whole scheme abhorrent. First of all the student has no regular personal contact with any single member of the staff. He can legitimately claim that no senior member of the university knows him well and that there is no direct channel between the individual undergraduate and the fountainheads of university authority and power. Of course I know that most universities, as a sop to the undergraduates, appoint what they are pleased to call "tutors" or sometimes "moral tutors" to look after the undergraduates' interests. The function of these moral tutors is quite different from the function of tutors at Oxford and Cambridge. The former are not responsible for the academic progress of their students but only for their general welfare while they are at university. They are supposed to be a combination of priest and general practitioner to whom the student in trouble can turn. The system is a farce. Very few of the moral tutors bother to make friends with the students before they run into trouble, and so it is not surprising that when a crisis comes the student does not naturally turn to his tutors for help. Almost all tutors are so busy with higher matters, such as research and university politics, that if he is lucky a student may see his man, shared with four or five others, at a sherry party once a term. This is hardly an inspiring way for those in authority to make the student feel that they care.

As effective means of imparting knowledge and stimulating thought, the practical classes and lectures which occupy virtually the whole of the students' time are often largely useless. In most practicals, the student walks into the laboratory, is faced with a few bits of apparatus and an instruction sheet and is asked to go through the motions of the prescribed experiment for the day. This experiment is usually some hoary old chestnut for which no original thought is required. This type of practical teaches nothing that could not be learned in ten minutes with a good textbook. It certainly does not stimulate real thought or give the student any worth-while concept of what research is like. It is difficult to understand quite what is the purpose behind such classes. Practice may be obtained in certain manual skills, but this is hardly a justifiable reason for forcing a student to spend whole days steaming in the laboratory. When he comes to tackle a genuine research problem he will find that many of the manual skills acquired in the undergraduate laboratory are of little use and that quite new ones are required. Even the undergraduate skills that are useful could be successfully acquired at the relevant times after about a week's intensive practice, without wasting the precious undergraduate years in this fatuous way. It is hard to avoid the suspicion that what inspires the designers of most practical courses is something like this. "Good heavens, the students have a free day to fill. How on earth can we keep them out of mischief?" There is a place for practical work and worth-while, thought-stimulating practical classes can be designed. But one such practical every month would do a lot more good than two of the other sort every week.

Much the same things may be said about lectures. Some university teachers like lecturing, take an interest in it and make the lecture an excellent method of teaching. Yet there are many dons who loathe lecturing and cannot be bothered to make a proper job of it. As a result, many lectures are useless and the undergraduate will often learn more from sitting for five minutes with a good book than in an hour listening to a poor lecture. Most students are fully aware of the poor quality of the tuition which they receive. Unfortunately they are usually prevented from showing their disapproval and shaming the lecturer by the size of his audience because they are forced to attend. If they fail to put in an appearance at a high proportion of lectures they may be required to take the course again before they are allowed to sit an examination. This is a barbarous practice. In a university neither lectures nor practicals should be compulsory. Compulsion merely isolates the dons from the silent disapproval of those they are trying to teach and enables them to go on their way without realising how perilously tenuous their relationships with undergraduates have become.

Another defect of modern universities is that most students have no time not taken up by formal tuition except in the evenings and at weekends. Many science courses are so organised that even in the evenings students are required to spend long hours neatly writing up the results obtained in the practical class of the day. Yet universities are supposed to be places where young people are encouraged to *think* and to use their own minds. Especially in the case of scientists, one might well ask, when? They are so preoccupied with soaking up information from lectures and with carrying out and writing up practical work that they never get the chance to think for themselves. I believe that this crazy situation accounts in large part for the oft-heard complaint that science students are dull, ox-like and wooden-eyed. They do not get the chance to be anything else.

The situation would not be so disastrous if all the compulsory courses which students must attend were worth-while. But many of them are worthless. Students listening to lectures spend half the time scribbling incomprehensible notes which they will never read and half the time wondering what the pretty girl in the second row is like in bed. It is not that they do not want to learn, but that the lecturers are so bad that they cannot learn even if they want to. It has often been suggested that all university lecturers should have to take a course in teaching methods and lecture presentation. I cannot see university teachers ever submitting to this, but even if they did I do not believe that the situation would be radically improved. The defect is not just that the lecturers have not been taught how to teach. The defect is that many lecturers do not care a hoot about teaching and no course in the theory of education is going to alter that.

In practice, as far as good teaching is concerned, the actual method used, whether it be the lecture, the practical class, the seminar or the tutorial, is probably not the finally decisive factor. Ultimately only two things make teaching an effective method of communication. The first of these is the motivation of the student and the second is the motivation of the teacher. If a student genuinely wants to learn and a teacher genuinely wants to teach, then sparks will fly and the school or university will be alive, even if the physical surroundings are appalling,

117

even if no visual aids are present and even if no teaching machines and other modern implements are available. But if either side is apathetic, the results must be disastrous for both, even if the buildings are splendid and even if no expense has been spared.

Anyone who has been both student and don in almost any university within the past few years should be able to tell why the present situation in the universities is so disastrous. There are inevitably a few rogue students, but I believe that nine out of ten (or more) of them come up to university desperately anxious to learn and to make the best of their three or four years. On the other side, there are undoubtedly some inspiring dons, but the majority of them find that teaching is an almost insufferable bore from which one must escape as soon as possible in order to get on with the truly important part of life, research. One might agree if university research were always brilliant and creative and if the frontiers of knowledge were being pushed back rapidly. But much of such research is mere hack work and many university lecturers know it, although they would never dream of admitting so to an outsider. One must be sympathetic because it is true that a man gets on in the academic world because of what he publishes and not because of how well he teaches. It is also true that the quantity of work published is usually more important than its quality, because most members of appointments committees will merely look at the list and will not bother to read any of the papers. Promotion in academic circles goes by the pound of print. In fact, many lecturers regard teaching ability as a negative asset, because if one is good at it and likes it one is inevitably asked to do more of it and less time is spent on research. I happen to think that they are not necessarily correct in this because it has often seemed to me that the people doing the most stimulating research are also often the best and most conscientious teachers. But my opinion is irrelevant because most university dons do believe that it is only research which matters and that teaching does conflict with research.

This lack of concern about teaching is, it seems to me, the fundamental cause of university unrest and the underlying reason why students are so unsatisfied. It makes the moderate student susceptible to the arguments of the few extremists who want only to destroy. But unfortunately I think that students, too, have got hold of quite the wrong end of the stick when they exert pressure for more students on university committees and for changes in the content of courses. Whatever is done along these lines can make only a marginal difference in a university atmosphere, because the motives which drive the acadamic staff will not be changed. If some brilliant student could produce a device whereby the tyranny of research could be removed and whereby dons would be forced to spend more time in meeting students face to face, then I think there might be a real chance of getting something done. But unless this happens, and I cannot honestly see it coming about, I for one am very pessimistic about the effects of any paper reforms which might be won.

At this point it is just worth mentioning the recent report of Mr. Aubrey Jones, formerly of the British Prices and Incomes Board, on the pay of university teachers. In this report, apart from commenting on the validity of the teachers' pay claim, he made some remarks on the way in which university salaries were determined. He suggested that too much time was devoted to research and too little to teaching. He also suggested that students should play a role in assessing the effectiveness of their teachers and in consequence the pay of those teachers. I personally think that he was unwise to make these comments and that his suggestion of student assessment was mildly ridiculous. It would quite clearly play into the hands of the flamboyant publicity-seeking don. However, although out of place and ridiculous, judging by the retorts of the dons which were equally out of place and often much more ridiculous, the comments of Mr. Aubrey Jones hit home. They were a warning that unless the universities put their own houses in order and do something about the abuses within them, other outside bodies, qualified or not, will be tempted to step in and force their hand. The reason for the temptation is clear. In my own subject, which I do not think is very different from any other in this respect, exceedingly few dons spend more than eight hours per week during term time in teaching. Many spend much less than this, and many teach for only two out of the three academic terms. If the teaching time is averaged out over the working year it comes to approximately the equivalent of rather less than one half day per week. For someone whose post is officially labelled as a university lecturer this is not very much. And yet to judge from the clamour which greeted the suggestion that more time should be spent on teaching, Mr. Aubrey Jones might have been asking for the moon. There was the usual pious nonsense about no notice being taken of the preparation required for teaching. A great deal of preparation is certainly needed the first time one gives a lecture course, but after that only the very pedantic or the very conscientious really devote much time to it. There was also a great deal said about the basis of university life being the advancement of knowledge through research. Without this continual renewal teaching would suffer. This may be true, but I cannot honestly imagine any great disasters if most of my colleagues spent an average of one day per week in teaching over the whole year instead of half a day. In fact I cannot think of anyone whose research is so vital and stimulating that if this were to happen the human race would suffer a severe setback. Nor do I believe that the teaching would be any less effective than it is in the dismal present. In my experience, I am afraid that the loud protestations of the academics just did not ring true. The motives behind them were, I suspect, either fury at the refusal of the board to recommend a bigger salary increase or horror at the suggestion that people who call themselves university teachers should actually spend just a little more time teaching. What I am afraid of is that, unless the universities react more constructively to constructive criticism, someone with more power than Mr. Jones will begin to lay into them with a heavier hand.

Naturally, being an Oxford graduate, I believe that the Oxford tutorial system at its best represents the ideal university education. The system has two levels, the colleges and the university. Every undergraduate receives tuition at both these levels. The university organises lecture and practical courses which are for the most part voluntary – although in recent years there has been an increasing tendency to make practical classes for scientists compulsory, a sure sign that those classes are becoming out of date and failing in their appeal to the undergraduate. But the centre of the Oxford system is the tutorial, which is organised by the senior members of the colleges known as Fellows, or tutors. These tutors are quite different in function from those called tutors at other universities. Once every week each undergraduate spends at least an hour with his tutor discussing an essay which the undergraduate has written. Sometimes the students come in pairs and occasionally in threes, but most often they are alone. Both don and undergraduate have to stay awake and pay attention to one another. They cannot sleep while someone else does the talking. The student is forced to learn to write and think for himself. He has to sit face to face with his tutor and he cannot snooze gently at the back of the lecture theatre, nominally fulfilling his obligations yet never once being forced to use his mind properly. The system has advantages for undergraduates at all levels of intelligence. Those who are not bright can ask elementary questions and receive elementary answers without fear of ridicule from other students. Those who are clever have the stimulus of face to face fencing with an equally clever but more experienced mind. Whatever the student's ability, he can never feel that there is no one in authority who knows him well. Through his tutor he always has an on the whole sympathetic channel to higher authority.

The Oxford system undoubtedly has advantages for the dons as well. They are able to live in one of the most pleasant cities in the world and can enjoy the stream of academically distinguished visitors who come to pay homage. The advantages of being a College Fellow in terms of good accommodation, good food and good company are enormous and are not to be measured in cash terms. Many Fellows find the life so satisfying that they have no desire to become Professors and Vice-Chancellors. Because early in life they achieve all they desire academically they do not feel the same pressure to pour out research as do the staff of most other universities. Indeed, they have at times been criticised for failing to be as productive as might be expected from their privileged position.

121

On the other hand, there are real disadvantages to being an Oxford don. The first is that a College Fellow is responsible for organising the tuition which an undergraduate receives throughout his course. He may send a student to other teachers for a term here and a term there but ultimately their success or failure is his responsibility. If students from his college repeatedly do well, his colleagues know where to bestow the praise. Equally, if the undergraduates from one college repeatedly do badly it is not only the students who are blamed. This is a real check on the effectiveness of teaching and is a real stimulus to the teacher. It is something which is completely lacking in most other universities where the dons have power without responsibility. All the students are taught by all the lecturers and so when the student fails it is no one's fault but his own. It is a comfortable system if you are a don but an unhappy one if you are a student. An introduction of the Oxbridge concept of the responsibility of a tutor for his pupils throughout their university career would, I think, be a real prize for student agitators in other universities to gain. But they should be warned that any such proposal might be fiercely resisted by the staff.

The second major disadvantage of being an Oxford don is that most of them spend between twelve and twenty hours a week in teaching. It is enough to make any non-Oxbridge research-minded lecturer run screaming from the dreaming spires in fright.

The Nature of Examinations

Finally in this chapter it would be appropriate, perhaps, to make a few comments and suggestions on the types of examinations which are set. As I discussed earlier, despite their drawbacks it does seem to me that examinations are in the interest of both teacher and student because they give an assessment more objective than that provided by any other method. Again, if any sense is to be made of the discussion, examinations should be considered under the separate headings of general education and vocational training.

The purpose of general education is to teach the young mind

to use words and facts and figures logically in reasonable argument. In this form of training, memory plays a relatively small part and the student should not be required to regurgitate a large amount of factual material. In most real life situations, he will be able to look up the facts in works of reference and will not need to have them instantly available in his mind. I think that an examination at the stage of general education should consist of two parts. First, there should be an intelligence test unrelated to any particular subject that has been studied. Although, as has been demonstrated earlier, these tests are undoubtedly not absolutely reliable and never will be, they are improving all the time. Even the present relatively crude ones have been found to be more useful at predicting ability at a particular task than have many more subjective tests specifically designed to examine the ability to do that task. For example, during and after the second world war, candidates for officer training in the British Army had to go through a complex procedure lasting three days and involving many different forms of tests of personality and of mental and physical ability. One of the test items was an intelligence examination. This single, simple short test was found to be a more accurate pointer to success in the subsequent officer training course than all the other tests together. The example could be multiplied many times. So in spite of their imperfections there is ample evidence that intelligence tests are at least as good in predicting success as are other more usual examinations.

Secondly, despite the success of intelligence tests alone, I believe that there should be an examination which tests not only the ability to do fragmented problems, but the ability to use facts to build up a coherent argument. This part of the examination should test the student in the particular subject or subjects which he has been studying. Again there is no point in requiring massive factual regurgitation. There is no point in requiring a student to do almost every one of a large number of questions. The examination should not be an underhand attempt to show up gaps in knowledge: it should aim to give the student freedom to demonstrate what he can do. There should therefore be a very wide choice of questions with only two or three out of ten or fifteen needing to be answered. In this way the student will not be forced to spend his pre-examination days in frantic cramming, because that will not help him. What is being tested is his ability to use his mind and he can be reasonably certain of getting questions which he can answer. The examiner is interested in how the questions are answered, not in whether they are answered or not. (Incidentally, although this is a debatable point, I feel that very much more weighting should be given to a brilliant answer than to a bad one. The person who can write a brilliant answer must have a mind which is capable of brilliance, and a bad answer in the same paper cannot alter that fact.)

There is no point in passing or failing students who take examinations at this stage of general education. One cannot fail a course in mental training. One can only do better or worse than one's contemporaries. Therefore everyone who, for example, does a degree course should get a degree. However, every degree examination and not just the honours variety should be classified, so that both the student and those who will later be dealing with him have some rough idea as to his mental ability.

Examinations in vocational training should naturally be quite different. I think that they should consist of direct tests of the ability to do a particular job. The examination for someone who is going to do scientific research should consist of scientific research aimed at a doctorate. The examinations for those going into the practical applied sciences, such as medicine or engineering, should be more conventional. There is a minimum number of facts which the doctor must know otherwise he will be lethal. I therefore would conceive of a vocational examination as consisting of two parts. One would consist of a wide choice of essay questions to check that the student understands the principles of the subject. The second would be a purely factual examination with a required pass mark of 80 per cent or so, merely to protect the public. The latter type of examination has a strictly limited educational function. Its main purpose is public protection and it has little else to recommend it. It is therefore a pity that, particularly in science, it is this type of heavily factual examination that has been almost universally applied at earlier stages where it really has no place (for example the ordinary and advanced levels of the British General Certificate of Education). Much criticism of examinations is effective and valid because such heavily factual examinations are employed at levels at which the main purpose of education is to train the mind. Their only proper use is in the testing of vocational knowledge, and if they were confined to that much of the sting would be taken out of the criticism of the examination system as a means for assessing students. An enlightened form of examination can give much more reliable evidence of a person's ability than can any subjective method of personal assessment.

Chapter 8

Science and Government

The aim of science is so to elucidate the workings of the natural world that that world can be explained or at least described by scientific laws. Theoretically, when this stage is reached, natural events become not only explicable but also predictable. Especially in a democracy it is important that government should be able to understand and to predict the behaviour of the people. If a government fails to appreciate the ways in which the people will react to acts of government, that government is unlikely to remain in power for long no matter how high-principled it may be. All politicians therefore realise that there is little point in having high motives if the reins are not in your hands. The attainment of power, and after that the maintenance of power, are the two main aims of those who aspire to govern. All else is subject to these.

History and Government

On the whole, governments are not very successful in predicting the responses of the governed to the acts of government. Indeed, to the casual observer it would seem that the results of any government action are almost invariably different from those which the rulers and their advisers hope to achieve. It is this failure to predict the ways in which people will behave which brings down governments. In a democracy a government is almost never broken because its policies succeed, no matter how distasteful those policies may be to some sections of the electorate. Governments almost invariably fall because their stated aims are not realised. And usually aims are not realised because ordinary people fail to behave as they are expected to.

As a consequence of this curious failure of professional competence, governments have increasingly asked outside experts to help them predict the future. The historians were perhaps the first in the field, and the belief that a good knowledge of ancient history would enable a man to become a better and more skilful member of the ruling class was one of the guiding principles of classical education. Unfortunately, in practice, a historical training has not been proved to be very useful in the prediction and control of people's behaviour. Its inadequacies have led to the famous pronouncements of practically-minded men: "The only thing that anyone learns from history is that no one learns anything from history" or, more pithily and crudely, "History is bunk".

There is a tendency to believe that the failure of prediction based on historical knowledge is not the fault of history itself but merely that of inadequate historians. It is suggested that the future might be predicted more reliably if only people would study history scientifically. Many university history departments contain eager young men anxious to apply what they are pleased to call scientific method to historical study. This attitude shows such a misunderstanding of science that few scientists bother to point out its defects. A knowledge of history is useless in predicting the acts of the people governed not because historical knowledge is at present incomplete and inadequate. It is useless because each historical event is starkly unique. Many hypotheses may be formulated about the causes behind any historical event, but a decision as to which, if any, is valid can be made only on grounds which a scientist would find laughable. It is true that the first stages of the scientific and of the historical study of a problem are similar. The scientist believes that natural events do not occur at random but that they are the manifestations of the operation of natural law. He believes in the logic of cause and effect and believes that natural events can be explained on this basis. Most historians also believe that historical events have a cause. Like the scientist, the historian believes in the ability of human reason to unravel the cause and effect problem. Like the scientist, he begins by collecting facts which bear on the issue. Like the scientist, he formulates a hypothesis when he feels that he has enough facts available. Like the scientist, he is prepared to modify his hypothesis in the light of further facts.

But there the similarity ends. The scientist is forced to submit his ideas to that peculiarly discriminating process, the experimental test. It may be worth briefly mentioning again the important features of a scientific experiment. The connection between hypothesis and experiment must be logical, tight and direct. Vague hypotheses cannot be experimentally tested. The factor under study and the variables which affect it must both be precisely measureable and accurately controllable by the experimenter. He must be able to do control experiments and repeat his work over and over again, as all too often the results of single experiments have been shown to be unreliable and false.

The historian never has to submit to such testing. Every now and again new facts may emerge which shed light on a historical event, but these can never be said to test a historical hypothesis in the way that a scientific hypothesis is taken through the fire. The unreliability of historical study is made clearly apparent by the radically different views which different historians may take of an event when the same information is open to all. Of course scientists, too, have radically different views of the same problem even when they all have the same information. The difference is that with science the hypothesis can be repeatedly subjected to experimental test until the truth emerges. With history the experimental test of a hypothesis is out of the question. And so the historian, with the certainty that he can rarely be absolutely refuted, is free to speculate *ad infinitum* and often *ad nauseam* as to the reasons why Hitler invaded Russia, why the Japanese attacked Pearl Harbour, why America became involved in Vietnam and why Russian forces entered Czechoslovakia. But for theoretical and not for practical reasons there can never be the "definitive historical study" of any one of these events.

If there can, therefore, be no certainty about the validity of historical explanations for past events, history must also be equally unreliable in its predictions of the future. History does not and cannot repeat itself except in the most superficial way. Any one historical event has a million possible causes and no two happenings can ever be quite the same. The fact that there may be some crude similarities in that perhaps the same nations are involved in the same corner of the world should not encourage one to believe that the same events will recur. The state of every nation this year is different from its state last year. Every person in power feels differently about a topic this year from the way he felt last year. The outsider, the journalist, the historian and the foreign diplomat can never truly understand how a decision is made. Even the insiders often show themselves to be somewhat bewildered about how it all happens.

If a politician or diplomat attempts to predict future events a deep intuitive understanding of the way in which human beings behave is likely to be much more helpful than an academic knowledge of history. The historian is naturally free to claim that the study of history is one of the best ways of achieving this deep understanding. Among others, Churchill believed that his knowledge of history was a major factor in enabling him to become a great leader. Unfortunately there is little evidence in favour of this view of the value of history. In my relatively short university career I have seen some of our most eminent historians involved in particularly inane squabbles. Their actions revealed clearly that their brilliant historical studies had given them virtually no insight into the way in which human beings behave. If the best historians do not gain this intuitive insight as a result of their work it is unlikely that such insight will accrue to those whose knowledge of history is less extensive. Historians, of course, are not the only academics who get involved in these foolish disputes. But the point is that historians do not get involved any less often than other members of university common rooms. This is certainly not a scientific proof of my contention that in practice a knowledge of history is little help in understanding human behaviour and no such proof is possible. Nevertheless, I think that almost anyone who has any experience of university life would agree that idiotic behaviour is unselective in its choice of those it afflicts. It is found in all faculties and none can claim immunity. Like kindness or like cruelty, this deep understanding of human nature which should be so important in government seems to appear at random in all sections of society, irrespective of class or educational background. It is as likely to be found in the roadmender as in the philosopher, and as yet there seem to be no rules which govern its development. It is certainly not acquired by any particular course of study. The great leader who has a deep knowledge of history may attribute his success to that. He forgets that others with an equally deep knowledge have made the wrong decisions because they lacked intuitive understanding. A knowledge of history, like the knowledge of any other subject, can be a great asset to a leader – but it does not make a leader.

Economics and Science

The failure of historians to provide a rational basis for political action has led to the elevation of other false gods, and of these economics is the chief. Despite the recent recurrent financial crises there are few politicians who are not convinced that all that is required to predict the economic future is a more perceptive understanding of economics. This ignores what ought to be self-evident, namely that in the last analysis economics is not a science but is merely a highly specialised form of history. An economic fact is a historical fact and not a scientific one. There is no science of economics and there never can be because the economic experiment is an impossibility. Experts may believe that a particular economic event was a consequence of this or that course of action. But they can no more prove it by experiment than can historians. The situation can never be recalled or repeated. Most important of all, control experiments cannot be done.

Since the concept of control experiments is all important yet is one which is quite unfamiliar to non-scientists, I should like to illustrate it from the sphere of medicine. Doctors have become used to the fact that many of the drugs which they think exert a particular action do not exert that action at all. The natural thing for a mother who has just produced a baby is to feed that baby with the milk being manufactured in her breasts. The production of that milk depends upon the actions of hormones released by the tiny pituitary gland at the base of the brain. When women turned from breast feeding in large numbers, the problem of breast engorgement became a serious one. The unsucked milk accumulated in the breasts, causing distension and often quite severe pain. Many varied and relatively ineffective methods such as binding the breasts and using hot compresses were tried. Then an artificial form of oestrogen (female sex hormone) known as stilboestrol became readily available in tablet form. It was known that in animals oestrogens can suppress the activity of the pituitary gland. It therefore seemed reasonable to suggest that oestrogen tablets given immediately after delivery might stop the pituitary gland pouring out its breast-stimulating hormones and might relieve the problem of breast engorgement. The tablets were tried and in most women they worked like magic. Their effectiveness was obvious, the logical approach had proved successful. For many years now stilboestrol tablets have been almost universally used in Europe and America for the suppression of lactation.

But recently doctors have begun to be concerned about the indiscriminate use of oestrogens. Admittedly very rarely, the drugs can cause clotting of the blood and serious illness. When this occurs as the result of taking tablets, no matter how rarely it happens, it is a tragedy. And so some doctors thought again about the suppression of lactation. They decided to do what in medicine is known as a controlled trial[9]. Women who did not want to breast feed were given tablets after delivery and were told that these were to stop the milk coming. Half the women were given oestrogen and half were given apparently identical tablets containing no active drug (placebo tablets). The women who received the placebo tablets thought that they were getting oestrogen just like the other women. The placebo tablets were as effective in stopping lactation as were the stilboestrol tablets. The fact of giving tablets to stop lactation seemed to be effective irrespective of what the tablets contained. The original reasoning had been logical but in spite of that had not been correct. This is an important moral which experimental testing of hypotheses forces to the attention of scientists again and again. Because an argument is logical, it is not necessarily correct. Correctness can be tested only by the use of control experiments. A control experiment is one in which the crucial factor under study is missing but in which all the other conditions of the experiment are fulfilled. In the case described the two groups of women were treated identically apart from the fact that one received stilboestrol while the other received placebo. Only then did it become apparent that the fact of giving a tablet could control lactation: it did not seem to matter whether the tablet contained stilboestrol or not.

In economics, control experiments are impossible. Suppose that a Professor Smith believes that excessive imports of chinchilla pelts caused the economic crisis of 1969. His belief can never be proved right or wrong. The crisis can never be repeated and it is impossible to set up a control situation in which all the elements are the same as before, apart from the fact that this time no imports of chinchilla pelts are allowed. This is the only way to demonstrate that the chinchillas really were the vital factor. Professor Smith can therefore happily assert that he is right. He may well be violently criticised by other experts but he can never be proved wrong. Scientific disputes can be resolved by controlled experiments, although naturally that does not prevent scientists from the fun of having slanging matches on the way. Slanging matches are the only arbiters of economic disputes.

If economics is only a special form of history, detailed study of European industrial expansion in the nineteenth-century, of the depression of the 1930s, of the relative economic failure of post-war Britain or of any other economic event is valueless in the prediction of the future. Since economic planning depends on economic predictions, this means that such planning is doomed to failure. Economic history is made in spite of government economic management and not because of it. I know about Sweden. Anyone who points to that country and says that her record proves the value of planning has failed to realise the necessity of controlled experiments before such an assertion can be proved. The question one must ask is: "Is it the economic planning which has really made the difference or is it the Swedish character, the freedom from war or the ready availability of certain raw materials?" There can never be a worthwhile answer because no control state identical to Sweden in every way, apart from the imposition of economic planning, could ever be devised. I suspect that even without any form of planning Sweden would not be very much less successful and that it is the Swedish character and the freedom from war which are the crucial factors. But I can never prove that I am right and nor can those who claim it is all due to economic planning. Apart from all this, it is dangerous for the advocates of planning to point to Sweden, for we can all point to societies much closer to home where planning has failed miserably. The advocates of planning always say that examples where planning has failed merely indicate bad planning. I suspect that what they actually indicate is that planning has little effect either way and that the real causes are quite different.

There is yet another factor which makes nonsense of hard and fast economic planning. This is that the state of the whole economic world can be altered in a moment by some political event over which the planners have no control. Events such as the closure of the Suez Canal, war in West Africa and dock strikes in Liverpool are not predictable in any reliable sense. That is quite different from saying that events of this sort cannot be predicted: they can and are by some shrewd observers. On the other hand, at the same time as one group of shrewd men is predicting the event, another equally shrewd group is predicting that it will not happen. Like all failures the latter tend to be forgotten. Newspaper prophecies are paraded before unfortunate governments after the disaster has happened and eloquent editorials say: "We told you so. If only you had listened to us all would have been well." They forget that at the time when it mattered other experienced men were saying quite the opposite. They also forget that in the last crisis their predictions were wrong. At the time when it really matters it is impossible to be sure about which particular piece of shrewd advice should be taken.

And so economics is a false god. It is history, not science, and it is as useless as history in predicting the future. The recent phenomenal expansion of economic studies is little but an encouragement of an unreliable academic indiscipline. This would not matter, it might even be a good thing, if the study were used as a training in reasoning which demonstrated the fallibility of the subject and of all related ones. Economics is useful as a training of the mind, but as a vocational study it is largely wasted. But the disturbing thing is that most economists, far from understanding the fundamental unreliabilities, actually seem to believe in economics. They explain its failures by saying that no one should have been so foolish as to follow that idiot Professor X's ideas. If only proper economic policies were followed, i.e. my own, then everything would be all right, they claim. In any case, even if we do have to admit that economics is a little unreliable at the moment, all will come right when we instal our bigger and better computers and when we can gather our facts more effectively. This argument does not appreciate that economics fails for theoretical as well as practical reasons. So-called economic laws are little more than the fevered dreams of economists. They have not been tested by the fire of the controlled and repeated experiment.

Sociology and Science

Sociology is another academic field which far too many people look upon as the hope of the future. It is not, and its faults are very similar to those of economics. A sociological fact is a historical fact and not a scientific one. The results of sociological research are historical results. They cannot be used to explain adequately the causes of past events or to predict accurately the course of the future. This is because the socio-logical experiment is an impossibility. Those who do not under-stand the proper meaning of the word experiment will no doubt say that this is nonsense. Look at this community or that community, they will say; surely they are sociological experi-ments. This is true only in the crudest, least useful and non-scientific sense. Suppose, for example, a slum community is taken from a great city and put into a carefully designed new town. That is not an experiment in the full meaning of the word because it is unrepeatable and because no control experiment has been performed. Suppose that the project seems to work and that the people are happy. Are we then justified in assuming that all slum communities in all similar new towns will be successful? Regrettably we are not. The slum community that is moved is unique. There is no other community like it in the world and there never can be. It is an individual, and just as two individuals may react quite differently when put into the same job, so two slum communities may react quite differently when put into the same new town. Secondly, every new town is unique, if not in the design of its streets and houses, in its position on the map. The slum community which responds well to one new town may do badly in another. Such mock scientific research as is done in sociology may in practice be extremely dangerous. It leads to general conclusions being drawn from unique, unrepeatable, individual cases. It tends to stifle flexibility and original thought about each particular situation. It leads to the Procrustean sin of forcing all issues which have one or two factors in common into a bed of the same type, irrespective of the multitude of differences which overwhelm the superficial similarities.

As with economics, the burgeoning of sociology as an academic discipline can be viewed with little but alarm. In 1967 in British universities there were about 185,000 students. Of these 34,700 were studying some aspect of sociology. Again this would not matter if people treated sociology as an exercise which demonstrated the unreliability and lack of wisdom of attempts to carry out what has hideously been called "social engineering" (another phrase which borrows the name of a solid, reliable, exact science and is used in the hope that some of the reliability may rub off on to sociology). If used as a demonstration of the uniqueness not only of the individual person but also of the individual group, sociology would be a most useful preparation for the complexities of life. But again, as with economists, many sociologists actually seem to believe in the reliability of what they are doing. They believe that all we need is more sociological information in order to plan with certainty our very own Great Society. It is a terrifying myth.

That special variety of sociologist, the educationalist, can be a particularly dangerous character unless he understands the fundamental uncertainties of what he is doing. This is partly because everyone wants the best for their children, and the man who shouts loudly enough that he knows and has proved what is best is likely to be believed. It is also partly because he is able to manipulate with relatively little protest young children whom, even if they can think for themselves, have on the whole been used to subordinating their opinions to those of the adult world. Parents who want the best and children who cannot protest provide fertile soil for the driving fanatic. Take, for example, the argument about grammar, secondary modern and comprehensive schools. The educationalist starts with two facts, that there are social divisions in Britain and that there are many children from working-class homes who are of high innate intelligence but who fail to develop their full potential. He then begins to hypothesise.

The first part of the hypothesis is that social division arises from and is perpetuated by the old division of state secondary education into grammar and secondary modern schools. This hypothesis is a sociological one, is unproven and unprovable. The second part is a corollary of the first, namely that social divisions will be eroded if we put everyone into comprehensive schools and abolish the other varieties. Again this is unprovable and one would have thought very questionable in view of the American experience where social divisions (and I am not thinking of black/white divisions) persist in spite of comprehensive education having been in operation for many years. The third part of the hypothesis is that all we have to do to make the best of a working-class child is to put him in the same school as the middle- and upper-class child. This of course ignores the influence of home life.

I cannot say that these hypothese are incorrect. All I can say is that it is impossible to prove in any simple way whether they are true or not. The educational experiment cannot be used in the argument as no such worth-while experiment can be carried out. Experimental schools are organised by unique individuals, and unique sets of parents send their unique children to them. Because Dartington or Summerhill or Kidbrooke or Eton or Marlborough work, it cannot be said that schools organised on similar lines but with different teachers, different children and different geographical locations will also work. The so-called educational experiment is first tried by an enthusiast utterly convinced of the rightness of what he is doing. The system may

work with him but it may well not work with the more run-of-
the-mill teachers and run-of-the-mill schools which make up the
greater part of the educational scene. No educational experiment can be valid in any reliable scientific sense because it
cannot be repeated and no controls are ever done.

This is not to say that all changes must be resisted because
they can never be proved to be right: but what it should make
apparent, however, is the fact that we should approach such
changes with caution and with an awareness of the possible
risks involved. We should not allow ourselves to be stampeded
in any one particular direction by claims which are incapable
of being verified. I am not saying that those who argue in
favour of comprehensive schools are wrong. All I am saying
is that there is no worth-while evidence on which the decision
can be taken. What is more, for theoretical and not only for
practical reasons, there cannot be any such evidence. There
cannot be any reliable evidence against the view, often described
as reactionary, that the destruction of the grammar schools
will do irreparable damage to the quality of school education
in Britain. But even though the evidence is and always will be
hopelessly inadequate, the British Government has decided
that British schools will become comprehensive and that the
grammar and secondary modern schools are to be abolished.
This central decision is being forced through with all possible
speed and any recalcitrant local authorities are being brought
rapidly to heel.

If the theorists who argue in favour of comprehensive
education are right, there can be no doubt that the result of
this change will be a great rejuvenation of Britain. But what
if they are wrong, what if much that was good has been destroyed to no avail? The results are horrible to contemplate. I
cannot oppose too strongly the central and universal imposition
of some current educationalists' dogma, not because I disagree with the dogma but because no one can know what the
results will be. Herein, surely, lies the strongest argument for a
greater measure of regional and local authority control. With
strong local control it is impossible to destroy over the whole
country something of whose virtue no one can be sure. If local

authorities had been allowed to decide what sorts of schools they would have in their own region, there would have grown up side by side in this country schools run according to the two opposing dogmas. Some authorities would have stuck to the old, some would have decided in favour of the new. In twenty or thirty years, when the dust had cleared a little, we might then have been able to make a marginally more rational decision as to which really was the better. But what has happened in practice? The one dogma, apparently terrified of genuine comparison with the other, has insisted that it can be given a fair trial only if the other dogma is totally eliminated. How dangerously close are we coming to the methods of the 1984 world.

If the results of central direction turn out to be right, the system will be much more successful and much more fair than if the two dogmas had been allowed to grow together. But if the central direction is wrong, the resulting inefficiency and disruption of our educational system will blight our future for many years to come. The end result will be far worse than if the two systems had been allowed to continue side by side. The argument for regional control in the realms of sociology and education is very similar to the argument for democracy. There can be no doubt that the efficient benevolent dictatorship is a far more effective form of government than is a democracy. Desirable things can be done quickly and the undesirable can rapidly be eliminated. But it is impossible to ensure that dictators will be either benevolent or efficient. Democracy is potentially neither so benevolent nor so effective as the régime of a good dictator, but it is infinitely better than the inefficiency and evil of a country where a bad dictator is lord. The central direction of social and educational policy may produce results which are much better than those which could be produced by regionalism. But they are more likely to be much worse. Muddling, bumbling control by the regions is unlikely to end up by successfully hitting the moon. Equally it is unlikely to drag the whole country into the ditch. We have yet to realise that efficiency is not the Supreme God and that the quest for it demands a price which may not always be worth paying.

Science and Government

Disillusioned by all these false gods, one at last turns to what seems to be our ultimate salvation, science and its offspring the computer. Here at least there seems to be a discipline reliable enough and solid enough to base future action upon it. First consider the computer. The computer is essentially a method of both storing and having readily at hand a vast amount of information which could not possibly be dealt with by more conventional methods. The amount of information which it can store and the speed with which it can manipulate that information are new phenomena which are going to change the face of the world in which we live. The computer is the ultimate hope of that army of academics which believes that it is merely lack of access to sufficient facts which has rendered both scientific and non-scientific disciplines unreliable. Their argument seems to be: "Economics, sociology, history and other subjects are clearly uncertain tools at the moment but that is because they are as yet immature disciplines. Once we develop computerised methods of collecting and correlating vast amounts of social, political and economic information, then these disciplines will become true sciences as reliable in their predictions as physics and chemistry are today." Professor Charles Tilly illustrated this state of mind well when in a book published in 1968[10] he bemoaned the fact that: "No country has a social accounting system allowing the quick, reliable detection of changes in organisational membership, kinship organisation, religious adherence or even occupational mobility." This is no doubt true, but I wonder whether Professor Tilly has thought out fully what he wants. What sort of country would it be if all this information were available to sociological researchers? Academics mislead themselves if they imagine that such detailed information about a population could be kept secret from politicians and other more sinister individuals who would only too dearly like to use it to manipulate people's lives. Once research workers have revealed a particular aspect of knowledge, that knowledge is no longer their personal property. They no longer have any control over what happens to it or over how it is used. The use of the atomic bomb in Japan should have taught that lesson. Research workers should therefore be more careful about the Pandora's boxes which they open to the world.

An important example of a sociological attempt to compile such a dossier on a particular group of individuals recently came to light. On November 28th, 1968, in the correspondence column of *The Times*, Mrs. Sally Oppenheim revealed that every new undergraduate at the University of Sussex is asked to fill in a 12-page questionnaire. The questions request information on the marital status of the student, on his parents, on the schools which he and his parents attended, on the size and status of his family home, on the religious and moral views of the under-graduate and his parents, on the newspapers and books which are read and on political affiliations. A similar questionnaire is again completed in the student's final year. The aim of the survey is to find out how the university alters the outlook of the student, and this information will be used in the future planning of the university.

Those who were carrying out the research put forward the usual defences. Undergraduates were identified by code number and not by name, there was no compulsion to fill in the form, the students were told that no one but the research team would have access to the questionnaires and that absolute secrecy would be maintained. I have no doubts about the integrity of the research workers but I would call into question their wisdom. They cannot guarantee their claim that the information will never be used except in the way in which they say it will be used.

If sociology and economics are so unreliable, what then does science, solid reliable classical science have to offer? Does it really offer any more hope that we may be able to plan the future more rationally? Unfortunately, I am afraid that when used in this way it is as unreliable as the other indisciplines. This is regrettably true for several reasons. First of all, anything which the physicist says about the future is valid only within the realm of physical laws known at the time. He cannot take into consideration aspects of natural physical law which are unknown. Nor can he take into account chemical and biological

laws of which he is unaware or which he does not fully understand. The same goes for the chemist, the biochemist, the biologist, the geologist, the engineer or any other form of scientist.
For example, the chemical engineer might say that a factory
produces so much carbon monoxide per year and that in ten
years time it will produce so much more provided that expansion
progresses at the same rate as at present. The limitations of his
prediction are obvious. It depends on an estimate of the expansion of the factory, and if that estimate is wrong the estimate of
carbon monoxide output must also be wrong. It also says
nothing about the effect of the noxious gas on the human
beings and on the countryside around the factory. The engineer
is not in a position even to begin to answer this question. He
and every other scientist must fail in their attempts to predict
the future. He cannot take into consideration aspects of natural
law which are not yet known. He cannot predict the effects
which his engineering will have on the environment as a whole.
He cannot know the rate at which his own special field will
develop. If the scientist's uncertainty about the future of the
physical world is so great, it is not surprising that he is certain
to be even less successful when he attempts to predict the
future of man.

Therefore the study of science and scientific methods demonstrates conclusively that more economics, more sociology, more
educational theory and even more science cannot be used to
predict the future. Governments should not be tricked into
believing that if they spend more money on these fields they
will govern much more successfully because they will be able to
predict more accurately the responses and the behaviour of the
people they rule. The current unprecedented multiplication of
economists and sociologists is no more likely to help than did
the multiplication of historians before them. Like history,
economics and sociology often seem to be used more for the
creation of new prejudices and the feeding of old ones than they
are to provide genuinely worth-while information. One example
of this is the fact that it is almost impossible for a Western
historian, economist or sociologist to communicate with a
Russian historian, economist or sociologist: they do not speak

the same language and there are no universally accepted basic principles. In contrast, Russian and Western scientists do communicate with one another freely and have no difficulties in understanding one another's form of communication. As far as prediction of the future is concerned, however, true science is as useless as the other subjects.

Scientific Lessons for Government

The lessons which government can learn from science do not come primarily from a study of the facts discovered by scientific research. They arise mainly from the scientific experience in devising reliable intellectual approaches to problems. Paradoxically, the study of the methods of science reveals that governmental policy and predictions are unlikely to be made more reliable or effective by the accumulation of vast amounts of information. The scientific lessons are largely negative ones. Science, by its use of the controlled experiment, has demonstrated the fallacy of the common belief that logical reasoning is necessarily correct reasoning. In order to be certain of being correct, a logical thought process must take into account every single factor which could influence a situation. No natural situation exists in which this is possible, and there are invariably hidden influences of which one is unaware. These hidden influences, like the effect on lactation of taking pills, can be revealed only by control experiments. In the great majority of situations in which government is interested, such controls cannot be carried out.

This negative lesson means that the dogmatic putting into practice of unproven and unproveable economic and sociological theories will almost certainly have consequences quite different from those which the advocates of the theories expect. Since such advocates naturally put their work in the best light, the consequences of their actions will usually be worse than they and others expect. It follows that if a government is going to introduce measures based on unproven dogma, such measures should be introduced cautiously and if possible on a regional basis. Only this caution will allow moderately fair comparison of the old and the new. Only this will prevent a system which is adequate being replaced by one which is a disaster.

If modern academic disciplines are so unhelpful, what characteristics should be possessed by the good politician, the good administrator or the good businessman? He must first understand with crystal clarity the unreliability of all information about the human condition and the danger of attempting to predict the future from a knowledge of the past. Secondly, he must above all be sensitive and humane. He must be able to appreciate the premises on which the other man's arguments are based. He must be able to understand intuitively how men will react to decisions which will change their lives. The great leader is not the first-class economist, sociologist or scientist. Being a scientist may help a little because the constant trampling by nature upon one's supposedly elegant hypotheses may perhaps induce a certain humility and consciousness of one's own limitations. The great leader is the one who understands that men want to do satisfying and worth-while jobs and who can provide that satisfaction. He is the one who understands that if men can be given satisfying work, and if they can be given confidence that what they are working for is worth-while, then economic management will not be needed. The economy will run itself. But once men fail to find their work satisfying, once they find that they are free to spend or to save less and less of the money they earn, then they will no longer be interested in working. If this happens a nation is heading for trouble and all the economic brilliance in the world will not stave off disaster. In order to put the financial affairs of Britain in order, leaders who understand men are required, not ones with firsts in economic theory. History, economics, sociology and science can help little. It is sensitivity and humanity which matter.

Facts of Science and Government Action

As I have emphasised, the fields within which science is a reliable tool are narrow indeed. The reliable help which a knowledge of scientific fact can give to government is therefore limited. There are few instances where scientifically-proven facts give a clear indication as to the action which a government should take. Even in these few cases the government action which is indicated, if a government is truly to act in the interests of the people governed, runs counter to powerful influences and so nothing is done. Two examples from medicine, the field with which I am most familiar, will suffice. No doubt other scientists working in other fields could provide equally striking instances.

The first is that of cigarette smoking. Approximately 30,000 people a year in Britain die of lung cancer. All but a tiny handful of these people are smokers. They often die in a particularly horrible way harrowing to themselves, to their relatives and to those who must treat them. They are often relatively young, and but for the cancer might expect to have years of useful life ahead of them. Again in Britain, every year about a quarter of a million people die as the result of disease of the heart and arteries. In this case, more factors such as obesity and lack of exercise are undoubtedly involved, but there is clear evidence that smoking greatly increases the risk of death from coronary thrombosis, particularly in the younger age groups. Therefore in Britain alone smoking kills at a prematurely early age at least 50,000 people and probably many more. The indications for government action on the basis of scientific fact could hardly be more clear. If Argentinian beef or French table wines were bringing into this country a disease which killed even ten people per year I have little doubt that the importation of these products would be totally banned. Yet we do nothing about something which is poisoning at least five thousand times that number.

Of course some of the reasons are not hard to see. A large part of government revenue comes from cigarette taxation. It is argued that if the advertising of cigarettes were totally banned, or that if their price were made prohibitively high by further punitive taxation, the total revenue would fall and the government would be in trouble. I have no doubt that the calculations of the fall in revenue are correct. But is there not something fundamentally rotten about a society which can survive only by peddling poison to its members? Surely we should be prepared to pay in other ways the blood price which will enable us to save the lives of 50,000 people in their prime every year. It seems not.

The refusal of government to do anything because of the fall in tax revenue is reinforced by another powerful factor. A high proportion of the population, including many members of the government and civil service, is itself addicted to the poison. No addict voluntarily cuts off his drug supply. One can sympathise with those young people who resent the laws against other drugs of addiction made and enforced by people already addicted to an equally pernicious poison. They might be wiser, however, if instead of agitating for a further spread of the drug-taking habit, they tried to persuade the government to take measures to reduce cigarette consumption.

The influence of alcohol upon driving is another example of how the findings of science fail to be properly used by government. There can be no doubt that there is no safe limit of alcohol consumption for those who are going to drive. Even the smallest quantities impair judgement. This is clearly recognised by airlines who absolutely forbid their pilots to take alcohol for several hours before flying. Yet legislation to ban completely drinking and driving could not be introduced because of the powerful social and commercial pressures, and the permissible level of alcohol in the blood was fixed at an unrealistically high level. Every young doctor who has worked in a casualty department knows that many drivers who come in for treatment and who are obviously the worse for alcohol manage to pass the breathalyser test. Most can tell horror stories like the following. Not long ago I was called out at 11 p.m. The casualty department was littered with bloody and broken bodies. A young man and a girl had been to a party. They had only had a few drinks and the breathalyser test on the young man who was driving was negative. Yet he had failed to stop at a halt sign when coming on to a major road and had smashed into a car carrying a mother, father, a boy of nine and a girl of six. The young man had a compound fracture of his leg and his clothes reeked of the beer which he had vomited up. The girl's face was cut to pieces. Her lower lip was cut right away and hung on by a mere sliver. She had to have over a hundred stitches and her once-pretty face will never be the same again. The father escaped with a broken arm and a few cuts. The mother was killed. The little boy was all right. The little girl had a huge hole in the side of her head. She survived but will be a lunatic cabbage for life. If a total ban on alcohol and driving stopped only one accident like that every year it would surely be worth-while.

It seems that those who loudly cry for government action based on scientific evidence do not realise what they are up against. If in cases like these, where the scientific evidence is so strong and so united, nothing can be done, how much less likely is it that anything will happen where smaller issues are concerned. It must regretfully be concluded that those politicians who call for more science and technology to be applied to government are paying only lip service to a fashionable concept. They are not prepared to act in situations where the evidence is clear but where powerful interests are involved. Until governments are prepared to act in cases like this it is hard to take seriously their pious statements about the role of science in decision making.

Science in Schools

Science can do little to help government to predict the way in which people will behave in the future. The range of subjects over which science is effective is narrow. Yet within this limited field, science is by far the most powerful tool we have for tackling problems. Therefore, in order to get the best out of science, government must understand it and must provide the best conditions for its healthy development. There are several ways in which government can help scientific advance, most notably in the financing of research, in education and in influencing the attitude of businessmen towards scientists.

Science must be taught in schools from an early stage if an adequate flow of scientific recruits is to be obtained. It should also be taught for another reason. An understanding of the scientific approach is a vital component of any educational system which hopes to fit a young person for life in the modern world. This is essential in order to protect people from having undue faith in the many fields of knowledge which attach the word "science" to themselves in the hope that their validity will then be more easily accepted. Political science and social science are two arch-examples of fields of study which are not scientific but which would like to have the reputation of being valid and reliable. Only if men understand what is implied by the scientific approach, only if they appreciate that any subject in which controlled repeatable experiments cannot be performed is not science, will they be in a position to repudiate those who try to seduce them with pseudo-scientific fantasy.

The chief difficulty facing school science is undoubtedly the miserable supply of teachers who have science degrees. For at least the next decade, it is impossible to see any means whereby the supply may be made adequate. It therefore follows that non-scientists must teach science with all the dangers that that implies. No subject is so dull to students as the one that the teacher does not fully understand. It is thus the urgent responsibility of the scientific community to make available to the teachers the necessary information in a lively, interesting but above all clear and understandable form. There is a tremendous need for stimulating elementary scientific textbooks. It is possible that at the moment the most important contribution which an eminent scientist can make to science is not a clever piece of research but a first-rate school textbook. He may then have far more influence by drawing young people into the profession than he could ever have merely by teaching and carrying out research in a university.

However, a supply of good books and teaching aids is no real substitute for the scientifically trained teacher. There is no more dismal spectacle than that of a teacher who does not fully understand mathematics or science trying to cram his own twisted ignorance into the heads of a class of equally uncomprehending children. This whole question of the supply of scientific teachers urgently needs evaluation. As industry pays higher and higher prizes to newly qualified scientists, so the number of good men prepared to go into teaching dwindles. There is one way in which the trend could be reversed. Teachers with science degrees should be paid more. This idea is anathema to the teaching profession, but it is anathema not because of any high-flown principle but because of that basest of human emotions which we all share, jealousy. The non-science teacher would probably complain, "I am not going to be paid extra for teaching my subject and so I am not going to let anyone else be paid more". In view of the motives of the opposition and of the national importance of a good supply of science teachers, a courageous government should be prepared to risk a political row over this matter. The opposition would be understandable, perhaps forgiveable, but certainly not justified.

Financial Support for University Science

This subject, too, presents thorny problems. Although some may claim that there is a shortage of university scientists, this is a relatively insignificant problem compared with that of finding scientists for schools and industry. At the moment I do not feel that the salaries of university scientists are a burning issue. I would doubt whether salary differentials are the most important factors which persuade scientists to go to the United States. The conditions under which research is carried out are much more significant.

One such condition which is so often little thought of by non-scientists is the excellence of the library facilities available. Access to a good library in my opinion is second only to the possession of a good mind among the factors which go towards the making of a successful scientist. Most British scientific libraries are dismal and, especially in the newer universities, they are often virtually non-existent. Most organisations which provide money for scientific research, whether public or private, are quite willing to spend extravagantly on expensive apparatus and buildings but are reluctant to provide finance for the purchasing of books. They tend to regard this as an internal university matter. This may be so, and the universities them-seleves are by no means guiltless in the way in which they spend the money available. I believe that conditions in this country for scientific research (and also for teaching and learning) would be much better if a greater proportion of the income available were spent on the purchase of books and periodicals.

The provision of money for equipment and technical assistance presents enormous difficulties, and I have great sympathy for the men who have to decide between rival claims. Universities are regrettably full of scientific second-raters who may never have had a good idea in their lives and who yet are consuming large amounts of government money in the form of research grants. Their research is all too often oh so pure and oh so useless. It is second rate in the sense that it consists of applying well-established techniques to problems only marginally different from those already solved. "Apply here, no original thought required" is a notice which might with justice be put up outside many university laboratories. Yet at the same time as this enormous financial waste is occurring, other genuinely worthwhile and important projects are held up for lack of cash. The problem facing any grant-giving body is to know which is worth-while and which is useless. No certain method for doing this can ever be devised, but I believe that the success rate would be much higher than it is now if finance was awarded not to projects but to men.

At the moment any scientist who wants money for research must submit details of his project and of his past research to the grant-giving body. He must describe in considerable detail what he intends to do and he must give good reasons for his belief in his own ability to do it. Most scientists loathe this project system for several good reasons. In the first place it makes logical nonsense, as the poor man is expected to say what he will find before the research is done. Often initial experiments reveal that the whole approach must be quite different than expected. Quite different equipment may be required, and before he can proceed further the scientist may have to waste time going again through the lengthy procedure of obtaining a grant. Secondly, untried hypotheses must be revealed to men who will be doing nothing but looking for loopholes. If it is remembered how often new hypotheses are out of tune with apparently well-established facts, it is obvious how easy destructive criticism is likely to be before the crucial experiments have been performed. Thirdly, it is usually assumed that the only people who can assess the project properly are those already working in the field and the grant-giving body may send details of the project to such people for their comments. Scientists are no better than they ought to be. An older man who sees a young man coming along to challenge his cherished ideas may find all sorts of good and honest reasons for saying that, while the young man is undoubtedly very able, it would perhaps not be the right time to provide money for this particular scheme. Perhaps more disturbing is the fact that with the best will in the world, a referee who has gained privileged access to a tentative and unpublished idea which throws light on his own work cannot forget that idea. Consciously or unconsciously he may incorporate it into his own research and gain an unfair advantage over the man whose dream it originally was.

I believe that the following scheme would avoid many of the inconsistencies and inadequacies of our present methods and would overnight halt the drain of the brightest scientists from any country which put it into operation. A carefully chosen committee should be appointed consisting of men noted for their brilliance in research rather than for their administrative ability. Ideally these men should have passed the stage when they might be worried about competition from other scientists. Each year the committee would review carefully the work done by younger scientists both in the universities and in industry and should pick out those who seem to be the most gifted and the most dynamic. Most scientists who are going to be really good have revealed this by the time they are 30 or perhaps by 35 in biology and medicine. These talented men should then be informed that for the next ten years a certain sum of money would automatically be made available for their use. In order to obtain this money they would simply have to ask for it without having to go through the process of submitting projects and fighting for every penny. They would have to produce a report on the use of the money only at the end of the ten-year period and not at yearly or six-monthly intervals. The correct size of the sum made available would have to be decided by trial and error: it might vary from subject to subject. Perhaps £5,000 per year might be suitable initially, making the cost of ten-year research support £50,000. Each scientist would thus know that without questions being asked he could spend £50,000 over a period of ten years. The system would be flexible so that a man who wanted to spend £20,000 or more in one year could do so, knowing that he would have less to spend in the future. On the other hand, someone who wanted to spend very little in one year could also do so knowing that the extra money would be available in the future and would not be lost if it were not used up. Incidentally, this is a particularly unsatisfactory feature of many university and other accounting systems for grants to scientific laboratories. Each department receives an annual grant of so much and any money not spent at the end of the financial year must be returned to the university. No departmental head with any sense is likely to return any money even if he does not want to spend it all in that year. He knows that such is the administrative mind that, when his grant position is reviewed, if he has not used all his money the size of his grant is likely to be reduced. Many heads of department would like to save money for their annual allocation in order to put the savings from several years towards some purchase which is too large to come out of a normal annual sum. The system in many universities actively discourages such thrifty planning and forces departments to use up their money on equipment which they do not really want and may never use.

Financial control over such a scheme would also be relatively flexible. The scientist would have to produce evidence to account for 90 per cent of the money spent. If he were required to account for 100 per cent he would end up wasting his time by trying to make purchases of rubbers and pencil sharpeners add up to the correct figure. The university or industry employing the scientist would still have to provide for the scientist the same amount of money given to other men of similar status. The government grant would be supplemental to such sums and in no sense a replacement for them.

Some may be appalled by the lack of tight control in such a system. I do not think that they need worry. A few rogues would probably be chosen but in most cases money spent in this way would produce a phenomenal return in terms of scientific advance. There would be little need to worry about the money being well spent. The able young scientist is only too eager to drive himself in his research laboratory. As I mentioned in an earlier chapter, the difficulty for his family and friends is usually not to encourage him to get down to it but to persuade him to take an occasional rest. To encourage those scientists who might tend to flag, there would be the spur of knowing that the work done would be reviewed at the end of ten years. If the first ten years had been fruitful, the scientist would be appointed for another ten. Except in rare cases that would be the end as after twenty-five or more years in research, relatively few scientists remain very active. Those who do go on are usually in positions of scientific power which enable them to tap other financial sources relatively freely. The cost of putting the idea into practice would be relatively small. For £7·5 million per year, very little more than the amount which the British Government spends in subsidising military bands, 1,500 scientists could be supported in the way I suggest. But I believe that such would be the rewards in terms of scientific advance and in terms of keeping good people in this country that £50 million per year would not be too much to lay out. Present methods of research finance would be continued undisturbed and all scientists would be free to apply for project grants in the usual way. But the very best would in large part be freed from the worst aspects of this ineffective and time-consuming procedure.

An important feature of the proposal is that it should be open as freely to young scientists working in industry and in the applied sciences as to those working in pure science in the universities. Industrial scientists are often frustrated by lack of funds to follow up their own personal ideas. Naturally enough, both public and private grant-giving bodies feel that in most cases it is the firm itself which should provide the finance for the research carried out by its employees. Also, naturally enough, and for the most part quite rightly, firms tend to be reluctant to finance research which is not directly and obviously related to some immediate company aim. They might well permit such research to go on in their laboratories if they did not have to pay for it. I think, therefore, that the prospect of government awards without strings being made to industrial scientists would greatly increase the scientific appeal of industry. It might also stimulate the more enlightened firms to take more interest in the research done by their young men in the hope that they might be able to win government awards.

Much government money is wasted in the financial support of doctorate students in the universities. It is misleading to think, as many of the students themselves appear to do, that the cost of supporting these men is only the £500 or so which is paid out to them every year in the form of a living grant. They cost money in less visible ways, in the provision of laboratory space and equipment and so on. Some of these young men would like to work in industry from the time of graduation, but they are deterred from doing this immediately after their first degree. This is because in most cases they will not then be able to obtain a doctorate from their own university because the university insists on residence during the doctorate studies. Many other young men will enter industry after obtaining their doctorates either because they fail to find a suitable academic position or because they do not enjoy life in the rarefied academic world. Yet, at the moment, both these categories of men are forced to spend three years doing research which is almost always both trivial and useless. Often they act as little more than intelligent but underpaid laboratory assistants for their supervisors. This situation could be altered to the mutual advantage of everyone by the introduction of two relatively simple measures. Firstly, universities should award doctorates on the basis of the calibre of research performed, irrespective of whether the work was done unsupervised in industry or in a university laboratory supervised by a university don. Secondly, industry should pay such doctorate students in the region of £1,000 to £1,200 per year for their services. Both industry and the post-graduate student would be getting a good bargain. The student would be paid far more than if he stayed on at university to do his doctorate research. The firm would get applied research done at a relatively cheap rate. They would be in an excellent position to interest the young man in an industrial career after his doctorate and would also be able to assess his ability before employing him on a permanent basis. In this way industry would be in a strong position to compete with the universities for the best brains. For purely economic reasons, many clever young scientists not particularly committed to an academic career but who would otherwise drift into university life would instead go into industry and stay there.

Publication of Scientific Work

Here I should just like to touch on a problem which many young scientists face. It is certainly not the concern of government except in so far as it makes it difficult for the brightest young scientists to be recognised as early as possible. Science here must put its own house in order. The problem is the almost universal practice of putting at the head of every research paper the names of all the people who have had anything to do with the work, including those who have merely given advice and those in whose department the work has been done. Papers ostensibly written by single authors are very rare, and those written by two are only a little less uncommon. Most research articles have three or four names at the top and some have ten or even fifteen. Many unscrupulous supervisors and heads of departments insist on their name going on the top of every paper written by one of their juniors, even if their contribution to the research has been only token. Naturally enough, those reading the papers, particularly if they have not been initiated into the common discourtesies of scientific writing, assume that it was the senior man who had most of the ideas, even if much of the practical work was done by the junior. In many cases this is quite untrue, all of the ideas and most of the experimental work having come from the young man. I can think of several distinguished scientific reputations which have been cleverly built up in this way on the basis of other people's work. I am by no means suggesting that all senior scientists are as dishonest as this, but enough are to make the problem a real one. Most of the research work described in a paper, no matter how many names may stand at the head. is carried out by one or two of the men concerned. The others are just senior colleagues or associates who chipped in with a bit of advice here and there. Science, and in particular that part of the scientific world represented by the editors of technical journals, has it in its own hands to supply the remedy. I believe that all research papers should be headed by the names of the one or two men who actually did most of the work. If a senior colleague kept an eye on the research, his name should go in only after the phrase "supervised by". If other scientists helped in some minor way their contribution should be acknowledged by words such as "advice given by". In no case should the name of a head of department go on a paper in any form unless that head was genuinely associated with the research in some way. I suspect that if journal editors insisted on these conditions the impressive work outputs of some distinguished scientists would apparently be strikingly reduced and some high reputations might be severely dented. Science could do itself nothing but good if it were a little more honest in this respect.

Government Research Institutions

Government is not concerned only with research in the universities and in industry. It also finances many research institutions of one form or another. The research done in these places is almost always applied. However, it can be divided into two categories, that which is applied to a specific and clearly defined problem and that which is applied in only a vague way to some long-term aim. Good examples of the first type are research into better road surfaces, into the development of Concorde or into a particular type of disease. This sort of research is obviously important, but politicians must be made to realise that it is inevitably expensive. Furthermore, at the beginning of a project it is quite impossible for any man, whether or not he is aided by a computer, to estimate reliably the size of the expense. This is because it is quite impossible to know before the research is done what lines that research will follow. Politicians must decide whether or not a project is essential, and if it is, then it is foolish to criticise costing too closely, because that may lead to the whole project being abandoned just when success is near.

The reasons for the high cost of such applied research must be clearly understood. The scientist who is free to choose his own problem deliberately decides to tackle ones which he feels can be solved relatively easily by currently available methods or ones for which new methods will be not too difficult to develop. The scientist, asked by a politician or civil servant or business-man with no real understanding of the matter to work on a specific problem, may be faced with the necessity of doing an enormous amount of preliminary research before the real issue can be tackled. For example, in order to put a man on the moon, a vast quantity of preliminary work must be carried out on electronics, on rocket propellents, on the stresses with which the human body can cope and on a myriad other things. The only answer to this difficulty is for the politician to be brutally honestly informed beforehand about the nature of the problem so that he can have a crude idea of the real difficulties involved.

Another reason for the high cost of this type of project is one that is not often mentioned. This is that relatively few scientists of the highest calibre are prepared to be told precisely what sort of problems they should tackle. They regard the choosing of a problem as a fundamental scientific freedom which can normally be obtained only within a university atmosphere. Only if the reward for relinquishing this freedom is very attrac-tive indeed can they be persuaded to abandon it. In Britain at the moment the salaries of research scientists in both govern-ment institutions and in industry are similar to those paid by the universities as far as men in their 30's and 40's are concerned. A man would therefore be a fool to give up the freedom of university life for the regimentation and direction of other types of research organisation. Inevitably, this means that in Britain both government and industrial laboratories tend to have more than their share of second- and third-rate scientists. Also inevitably, these men work less effectively than those with genuinely brilliant creative brains. Projects thus take much longer than they ought to and are much more expensive. If the country is to survive and to maintain its present status, it is imperative that many more of the very clever should be attracted from the universities to work for industry and government. This can be done only by paying genuinely high salaries of £10,000 a year or more, accompanied by all the perquisites which would be given to a businessman of similar status. If a project is to cost millions of pounds, the employment of the brilliant, far from being a waste, will ensure that the work is completed as quickly and effectively as possible. By paying very high salaries a great deal of money may be saved.

The other type of government research can be said to be applied only in a vague sense. A research unit may be set up to study a subject which is clearly going to have practical implications in the future but which in the present is not of much use. Any scientist could pick out a number of such fields with ease. Obvious ones which have already attracted a great deal of attention and cash are nuclear and space research. Less obvious ones for which I should like to make a special plea are biology and oceanography. In neither of these areas is any other nation very far ahead of Britain. In neither area is research excessively expensive, especially compared with the appalling costs of nuclear and space research. The Americans can afford to spend more than fifty million pounds on one space shot which may go wrong. That amount of money spent by Britain on oceanography would give us a lead in that field so long that any other nation would find it very difficult to catch up.

I believe that it is the success or failure of research into biology and oceanography which will ultimately decide whether or not man is going to survive on this planet. Unfortunately, when we interfere with nature, all the golden rewards come rapidly but the evil tends to be a slow, long drawn out affair. When we put fertiliser on our fields, the startlingly good effect on plant growth is seen within a year or less. The slow choking and poisoning of our inland rivers and lakes which that fertiliser brings about comes much later. When we use chemicals to destroy the insects which damage our crops, the effect on agricultural production is immediate. The evil influence of the chemical on birds and animals, and ultimately on ourselves, comes much later. When we introduce to man the joys of nicotine or of food sweetened by cyclamate, the rewards are immediate and only after many years does the rottenness set in.

And then it is far too late. We need biologists desperately. We need them to build up as quickly as possible information about nature and about the delicate interactions between species which operate at present. We need them to test much more thoroughly than at present the effects of every new industrial process and every new chemical which we propose to release into the world. We want to find out about the cancer-promoting effects of cigarettes and oil and asbestos before and not after they are brought into widespread use. And yet the number of biologists who can do this sort of work is pitifully small and derisory sums are available for the support of biological research. But unless we increase our output of biological scientists very rapidly, and unless we take a great deal more notice of what they say, we are going to be in deep trouble indeed. I am fully aware that if all new industrial processes and chemicals had to be stringently tested for their biological safety, this would put a severe brake on what we are pleased to call progress. I am fully aware that we should have to forego many short-term advantages. I am aware that agricultural production and economic growth would proceed less rapidly. The population would inevitably suffer in the sense that expected rises in the standard of living might apparently take longer. But they would suffer as the man suffers who refuses to go into the sleasy whore-house. He misses the immediate pleasure: he also avoids the pox. And for the pox which we are at present inflicting upon the natural world there may be no cure. Compared to this the hydrogen bomb is a child's toy. The nuclear bomb is big, it is dramatic, it obviously hurts and it has no attractions whatsoever. It frightens people and makes them behave. The biological danger seems at the moment to be small. It is insidious, it cannot easily be seen by the biologically unsophisticated and it lures men on because of its dramatic short-term advantages. Yet ultimately it will bring death, as the life on our planet on which man depends is destroyed. It will bring death much more surely than the bomb. Unless we act now to stop this senseless destruction of our natural environment, a hundred, a thousand or perhaps ten thousand years hence our descendants will pay the ultimate penalty for our folly. But the reckoning seems such a long time away. This is the way the world ends, not with a dramatic bang as the nations blow themselves to pieces, not with a noble sigh as the sun gradually dims and fades away, but with a foul stench as man commits suicide by wallowing in his own waste.

On the positive side, in enhancing the quality of life, oceanography is as important as is biology in helping to prevent its destruction. Over three quarters of the earth's surface is covered by sea. Over 85 per cent of that sea is over one mile deep. Yet we know less about the sea that long one mile below us than we know about the moon a quarter of a million miles above. As yet the sea has proved too vast for man's spoliation. He has committed some acts of carnage. Some species of whale are almost extinct and some fishing grounds have been worked almost to destruction, but on the whole the ocean is virginal. It will not remain that for long. The interest of other nations is beginning to stir. Should not we British, who are surrounded by the sea and whose livelihood has always depended on it, be at the forefront of the field? If we do not work on the problems of oceanography now, other nations will, and we shall later on be forced to pay them for their knowledge. On the other hand, if we get in early on a big scale, in 20, 40 or 60 years' time we shall be able to export our goods and know-how. The reasons for this are twofold. First, as has been repeated *ad nauseam*, the world's population is growing far more rapidly than its food supply. On any realistic estimate it will be many years before contraceptive techniques which are simple, effective and acceptable to all can be applied on a scale which will stabilise that population at a constant level. At that stable level, if it is ever achieved, the world's population will be many times what it is now, and it is most unlikely that the land will be able to provide all the necessary food. We must therefore farm the sea. Primitive experiments have already begun but for the most part, as far as the sea is concerned, we are still operating a Stone Age food-hunting and food-gathering economy. We cannot afford to stay in this state for long, and before this century is out the world will be clamouring for the machinery of sea agriculture. I wonder if Britain will supply it.

The second good reason for supporting research into oceanography is that the solid and liquid mineral resources of the world are being used up at an appalling rate. For most of these minerals the known deposits are sufficient to last for many years at present rates of use and there are undoubtedly many other deposits not yet discovered. But this cannot circumvent the fact that there must be a finite limit to the size of the deposits of each type remaining. Nor can it disguise the fact that in most cases we are moving towards those limits at a dangerously high rate. Once land-based minerals are used up, only two other types will be left, those on and beneath the sea bed and those dissolved in the sea itself. The need for industrial processes for the extraction of these minerals may be many years away, but the exploitation of undersea gas and oil has revealed some of the possibilities. Again, the nation which begins work now will be in a strong position when the crises of mineral supply come later. Is it too much to ask that in just one or two things governments should look thirty or forty or even a hundred years ahead? "A week in politics is a long time" is one of the most pathetic and dangerous, yet regrettably true, comments ever made.

Scientific Advice for Governments

The problem of the way in which a government can acquire the necessary scientific information and advice is a very real one which as yet has received no satisfactory answer. The traditional solution is to employ advisers who visit Whitehall several times a year and expound their pet theories at great length. There are several drawbacks to this. The advisers are usually chosen for their administrative rather than for their scientific ability. They are inevitably the establishment figures of science, and the voice of the young and brilliant out-sider is rarely heard. For reasons simply of convenience, if for none more sinister, the advisers are drawn too heavily from London and the two ancient universities. Voices from more distant educational institutions are much less often heard. The advisers are often ineffective because they fail to understand government and the ways in which political decisions are made. Even when they are effective and when their advice is good, in the last analysis they are only advisers and there is no obligation for a government to act upon their advice.

Too, the adviser system often fails because it needs someone with a scientific background to appreciate scientific advice. Any doctor who has had to deal with highly intelligent but scientifically untrained patients knows how easy it is to parry and to answer awkward questions asked by those who know nothing about a subject. I am sure that this is true of all other scientific fields. The expert adviser can so "blind with science" the politician or civil servant that it is impossible to assess realistically the value of the advice. Only scientists can ask the nasty questions which should be asked and can avoid being taken in by slick answers. Only scientists can truly assess the worth of any scientific adviser. It is all too easy for the non-scientist to be taken in by the forceful and ruthless but inherently unsound scientific adviser. It therefore follows that the only real solution to the problem of providing effective scientific advice for governments is to have scientists actually in positions of genuine decision-making power, both in the civil service and as Members of Parliament. But there are very few MPs with science degrees, and I do not believe that there is a single one who has spent more than a very few years as a practising research scientist. Until this position changes we are not likely to get a sane scientific policy with sensible support given to research and where appropriate the results of scientific research being put into effect as quickly as possible. The situation is not entirely the fault of government. They often try hard to obtain sound scientific advice but they do not know where to turn. It is largely the fault of the scientific community for failing to produce men who are prepared to devote their lives and their knowledge to the service of government. Scientists must realise that if they are not prepared to take part in the democratic process they can hardly criticise the decisions which are taken. They must come out from their laboratories and become local councillors, civil servants and MPs. Only when scientists in government service are as common as lawyers can we expect government and science to serve one another in the most fruitful way.

But the scientific community is not entirely to blame, particularly as far as the Civil Service is concerned. This service has in recent years been making approving noises about the recruitment of scientists into administration. A close friend of mine with a top first-class honours degree, a doctorate and several years' practical research experience, recently joined a ministry very much concerned with scientific affairs. On the first morning one of the deputy heads of his ministry contacted him by telephone. "We wondered whether while you are working for us you would mind calling yourself *Mr.* and not *Doctor.* To have a Dr. X in the ministry might cause confusion. People might think that you were a *real* doctor and it also might cause ill feeling among those who do not have higher degrees. I also want to point out that while you are here you are not permitted to express any opinion on scientific matters. We have our outside advisers for that purpose and you will not be expected to question their advice." My friend formed on the first day an opinion of the Civil Service which persisted relatively unchanged for two years. During that time he was not on one single occasion asked his opinion on any scientific problem. His expensive education and his post-graduate experience were entirely wasted. At the end of the second year he could stand it no longer and with relief crawled back to his research and teaching at the university. And on that happy note which augurs so well for our future I will end this chapter.

Chapter 9

Conclusion

Science is the modern god. In a disturbingly large number of ways, the position of science in the mid-twentieth-century parallels that of religion in the mid-nineteenth. Twentieth-century politicians invoke science and justify their actions on the basis that they are done in its holy name. Twentieth-century scientists, like nineteenth-century theologians, make the wildest claims on behalf of their god, not realising the danger that if these claims are proved false their god may fall. Twentieth-century charlatans of a myriad varieties offer their panaceas for society and attempt to mislead the people by calling their misbegotten concepts scientific. And bewildered twentieth-century common men have a crude faith in their god which they do not care to have questioned too closely but which could be destroyed if it were demonstrated that their graven image has feet of clay.

The lack of reality and hard thinking is terrifying to behold. Science is without doubt the most powerful method of intellectual enquiry yet devised by man. But it is valid only within very strictly limited fields, those within which measurement is reliable and where repeatable controlled experiments are possible. It is not the highly systematic gathering of much information which makes a subject scientific, nor is it the formulation of plausible hypotheses on the basis of that information. A subject becomes scientific only when its hypotheses are subjected to the furnace of repeated and rigidly controlled experimental tests. Social science, economic science and political science are non-sense terms. We should have no more faith in the proposals for future action made by sociologists, by economists and by political theorists than we have in the explanations for past events advanced by historians. The attempt to trick the public into believing that these modern intellectual approaches to aspects of life have a true scientific validity can lead to nothing but disaster.

Science is God

When used properly to tackle a problem which falls within its horizons, science is a superb and incomparable tool. When used improperly, and when its discoveries are loosed upon the world by politicians and industrialists who do not fully understand them, science is a dangerous menace. It runs the risk of providing the means whereby the environment upon which we depend is either slowly poisoned or burned up in a spectacular flash. It would be better to stop scientific research altogether than to allow that to happen. However, only the naïve would imagine that the total cessation of research and technological advance is a practical measure for stopping the apparently inexorable movement towards our own destruction. We cannot expect the majority of politicians, industrialists and administrators to understand science, and I do not believe that the better education of such people offers any real hope. The only possible answer lies with the scientists themselves. They must abandon their privileged mistress role of advisers who do not have to carry the can. They must examine the implications of their own research more closely and must voluntarily discard the proposition that all research is legitimate and is justified by pure curiosity alone. They must shout loudly when charlatans attempt to usurp the word science in order to give more respectability to their quackery. They must take much more seriously their responsibilities in informing the public of the implications of what they are doing. They must insist on a much greater emphasis on biology and must allow biologists to investigate the effects on the environment of scientific and technical advances made in other fields. But above all they must be prepared to get their hands dirty in the process of government. I think it was the late Lord Florey who said that the laboratory which is clean and tidy is the one where nothing important is happening. Scientists must stop being so proud of being untainted by contact with industry and government. If they are untainted they are also certain to have no influence. Scientists must leave their laboratories and become themselves industrialists and administrators and politicians. Only when we have as many scientists as lawyers in public life is there any hope of a sane scientific policy. Only then shall we have any real chance of avoiding the plotting by accident of our own destruction.

References

1. *The Act of Creation*, Arthur Koestler, Hutchinson, 1964.
2. *The Art of the Soluble*, P. B. Medawar, Methuen, 1967.
3. *The Double Helix*, James Watson, Weidenfeld & Nicolson, 1968.
4. P. A. M. Dirac, *Scientific American*, May, 1963.
5. *Controversy in Internal Medicine*, edited by Ingelfinger, Saunders, 1966.
6. Enquiry into the Flow of Candidates in Science and Technology into Higher Education. (The Dainton Report) H.M. Stationery Office 1966.
7. *The Flow into Employment of Scientists, Engineers and Technologists* (The Swann Report). H.M. Stationery Office, 1968.
8. *The Employment of Highly Specialised Graduates*, M. C. McCarthy, H.M. Stationery Office, 1968.
9. Inhibition of lactation by oestrogens, S. J. Steele, British Medical Journal, *4*, 578, 1968.
10. Charles Tilly in *American Sociology*, edited by Talcott Parsons, Basic Books, New York, 1968.